The Ethology of Domestic Animals

An Introductory Text

To Ingvar Ekesbo

The authors would like to dedicate this book to a person whose devoted work in this area has contributed greatly to the progress of applied ethology. Long before it became widely accepted that understanding animal behaviour was an important aspect of understanding animal welfare, he actively promoted the development of applied ethology in veterinary medicine and animal science. The importance of his work for Swedish and European animal welfare development can not be overestimated, and this work continues actively.

The Ethology of Domestic Animals

An Introductory Text

Edited by
Per Jensen

Professor of Ethology
Swedish University of Agricultural Sciences
Department of Animal Environment and Health
Skara, Sweden

and

Division of Biology, IFM
University of Linköping
Linköping, Sweden

CABI *Publishing*

CABI is a trading name of CAB International

CABI Head Office
Nosworthy Way
Wallingford
Oxfordshire OX10 8DE
UK

Tel: +44 (0)1491 832111
Fax: +44 (0)1491 833508
Email: cabi@cabi.org
Web site: www.cabi.org

CABI North American Office
875 Massachusetts Avenue
7th Floor
Cambridge, MA 02139
USA

Tel: +1 617 395 4056
Fax: +1 617 354 6875
Email: cabi-nao@cabi.org

A catalogue record for this book is available from the British Library, London, UK

Library of Congress Cataloging-in-Publication data
The ethology of domestic animals : an introductory text / edited by Per Jensen.
p. cm.
Includes bibliographical references (p.).
ISBN 0-85199-602-7 (alk. paper)
1. Livestock – Behavior. 2. Domestic animals – Behavior.
I. Jensen, Per
SF756.7.E838 2002
599.15–dc21 2002000665

ISBN-13 978-0-85199-602-8
ISBN-10 0-85199-602-7

First published 2002
Reprinted 2003, 2005, 2007

Printed and bound in the UK by Biddles Ltd, King's Lynn

Contents

Contributors

Morten Bakken, *Agricultural University of Norway, Department of Animal Science, PO Box 5025, N-1432 Ås, Norway*
Email: morten.bakken@ihf.nlh.no

Bjarne O. Braastad, *Agricultural University of Norway, Department of Animal Science, PO Box 5025, N-1432 Ås, Norway*
Email: bjarne.braastad@ihf.nlh.no

Björn A. Forkman, *Royal Veterinary and Agricultural University, Department of Animal Science and Animal Health, Grønnegårdsvej 8, DK 1870 Frederiksberg C, Denmark*
Email: bjf@kvl.dk

David Fraser, *Animal Welfare Program, Faculty of Agricultural Sciences, University of British Columbia, 270-2357 Main Mall, Vancouver, BC, V6T 1Z4, Canada*
Email: fraserd@interchange.ubc.ca

Stephen J.G. Hall, *Lincolnshire School of Agriculture, University of Lincoln, Caythorpe Court, Grantham NG32 3EP, UK*
Email: sjghall@dmu.ac.uk

Per Jensen, *Swedish University of Agricultural Sciences, Department of Animal Environment and Health, PO Box 234, SE-532 23 Skara, Sweden and University of Linköping, Division of Biology, IFM, SE-581 83 Linköping, Sweden*
Email: Per.Jensen@ifm.liu.se

Linda Keeling, *Swedish University of Agricultural Sciences, Department of Animal Environment and Health, PO Box 234, SE-532 23 Skara, Sweden*
Email: linda.keeling@hmh.slu.se

Sue M. McDonnell, *Equine Behavior Program, University of Pennsylvania, School of Veterinary Medicine, New Bolton Center, Kennett Square, PA 19348, USA*
Email: suemcd@vet.upenn.edu

David B. Morton, *University of Birmingham, Birmingham B15 2TT, UK*
Email: d.b.morton@bham.ac.uk

S. Mark Rutter, *Institute of Grassland and Environmental Research, North Wyke, Okehampton, Devon EX20 2SB, UK*
Email: mark.rutter@bbsrc.ac.uk

Frederick Toates, *Open University, Milton Keynes MK7 6AA, UK*
Email: f.toates@open.ac.uk

Daniel M. Weary, *Animal Welfare Program, Faculty of Agricultural Sciences, University of British Columbia, 270-2357 Main Mall, Vancouver, BC, V6T 1Z4, Canada*
Email: dan.weary@ubc.ca

Preface

Modern farm environments differ strikingly from the natural habitats of the ancestors of today's farm animals. Also, the animals look different. However, just under the surface, surprisingly little has changed in the behaviour of domesticated animals. Given a chance, they still behave very much like their ancestors, and it is only possible to understand some apparently bizarre actions in the light of the ancestral history of the species. On the other hand, some of the behaviours we can observe in animals in a modern farm or in a laboratory are not part of the normal, species-specific behaviour at all. They may even indicate that the animal is under stress and that its welfare is poor. Distinguishing between these possibilities is one important goal for applied ethology.

This book aims at introducing the subject of ethology and its applications. The readers we have had in mind when writing the book are, for example, students of animal science or veterinary medicine, or biology students taking introductory courses in animal behaviour or applied zoology. We have assumed that the readers will have some fundamental knowledge of basic biology, for example, physiology and genetics, and probably of modern animal husbandry, but that this book will be one of their first texts in animal behaviour.

The authors of the book are all active in research, but they are also experienced teachers. This combination was a main factor in selecting the authors. Hopefully, this will mean that the text has a suitable level and content. In order to make the text more easy to read, the authors have reduced the references to a minimum, and mainly refer to central publications which will provide a basis for further studies. Hence, not all cited experimental data are explicitly referenced, but can mostly be located via certain of the other publications in the reference lists.

Per Jensen
Skara, October 2001

Part A Basic elements of animal behaviour

Editor's Introduction

The first six chapters of this book introduce the basic concepts and the central subject matters required for a firm understanding of the biological basis of animal behaviour. These chapters are not necessarily tied to domesticated animals, since the basic matters treated are the same for most species. The first chapter provides a historical background, which may help in understanding the questions that occupy contemporary ethology and its applied branches. In the second chapter, we approach the important issues of if and how behaviour is controlled by genes (the nature–nurture debate) and also what this means for understanding behavioural evolution. This chapter also describes the process of domestication, which is essential for understanding present-day domestic animals. The third chapter goes to some depth in describing how observable behaviour is a result of processes in the brain, which in turn are affected by various physiological processes throughout the body. The concept of motivation has a long history in ethology, and has proven to be essential for understanding the needs of animals in captivity; the concept receives a detailed treatment in this chapter.

In the fourth chapter, we move our emphasis from mainly genetically and physiologically controlled mechanisms (i.e. internal mechanisms) to mechanisms of behavioural change caused mainly by varying external conditions – summarized as learning. Learning is central in adaptation to any environment, and this chapter therefore provides some of the bases for understanding how domestic animals can function in various artificial conditions. The chapter also approaches the question of animal cognition, essential for judging their perception of the world and their welfare. The fifth chapter moves to a more evolutionary and ecological approach to behaviour. Social and reproductive behaviour are important elements of applied ethology, since animals are normally kept in groups, and the animals are expected to reproduce.

The sixth and last chapter in Part A attempts to bridge the gap between basic and applied ethology. It raises some of the central issues of contemporary applied ethology, such as how the behaviour of an animal can be used to assess its welfare and whether it is under stress. It also gives a broad and general outline of the most common and important behavioural disorders seen in captive environments.

1 The Study of Animal Behaviour and its Applications

Per Jensen

Introduction

Ethology is the science whereby we study animal behaviour, its causation and its biological function. But what is behaviour? If we spend a few minutes thinking about this, a number of answers may pop up which together illustrate the complexity of the subject. In its simplest form, behaviour may be a series of muscle contractions, perhaps performed in clear response to a specific stimulus, such as in the case of a reflex. However, at the other extreme, we find enormously complex activities, such as birds migrating across the world, continuously assessing their direction and position with the help of various cues from stars, landmarks and geomagneticism. It may not be obvious which stimuli actually trigger the onset of this behaviour. Indeed, a bird kept in a cage in a windowless room with constant light will show strong attempts to escape and move towards the south at the appropriate time, without any apparent external cues at all.

We would use the word behaviour for both these extremes, and for many other activities in between in complexity. It will include all types of activities that animals engage in, such as locomotion, grooming, reproduction, caring for young, communication, etc. Behaviour may involve one individual reacting to a stimulus or a physiological change, but may also involve two individuals, each responding to the activities of the other. And why stop there? We would also call it behaviour when animals in a herd or an aggregation coordinate their activities or compete for resources with one another. No wonder ethology is such a complex science, when the phenomena we study are so disparate.

But how did it all begin, and how has ethology developed into the science it is today? This chapter will provide a brief overview of some landmarks in its history, and of the various fields into which the science has branched over the past decades. The field that interests us most in this book is, of course, the applications of ethology to the study of animals utilized by humans.

The History of Animal Behaviour Studies

No doubt, knowledge of animal behaviour must have been critical for the survival of early *Homo sapiens*. How could you construct a trap, or kill dangerous prey weighing several times your own weight, unless you had a genuine feeling for animal behaviour? So it should not be of any surprise that the earliest 'documents' available from humans – cave paintings up to 30,000 years old – are dominated by pictures of animals in various situations. Written, systematic observations and ideas about animal behaviour were published by Aristotle more than 300 years BC (Thorpe, 1979).

One of the first to write about animal behaviour in a modern fashion was the British zoologist John Ray. In 1676 he published a scientific text on the study of 'instinctive behaviour' in birds. He was astonished by the fact that birds, removed from their nests as young, would still build species-typical nests when adult. Ray was unable to explain the phenomenon, but noted the fact that very complex behaviour could develop without learning or practice (Fig. 1.1). Almost 100 years later, French naturalists had an important influence on the development of the science. For example, Charles Georges Leroy, who was not actually a formally trained zoologist, published a book on intelligence and adaptation in animals. Leroy heavily criticized those philosophers who spent their time indoors, thinking about the world, rather than observing animals in their natural environments. Only by doing this, he argued, would it be possible to fully appreciate the adaptive capacity and flexibility in the behaviour of animals (Thorpe, 1979).

Fig. 1.1. A sow needs no prior experience to be able to construct an elaborate nest before farrowing. 'Instinctive behaviour' such as this fascinated early behavioural researchers.

Another 100 years on, two important scientists deserve to be mentioned. The first is the British biologist Douglas Spalding, who published a series of papers on the relationship between instinct and experience. Spalding was way ahead of his time in experimental approaches. For example, he hatched eggs from hens by using the heat from a steaming kettle, in order to examine the development of the visual and acoustic senses without the influence of a mother hen (Thorpe, 1979). The second important scientist is no less than Charles Darwin.

Darwin has probably had the most significant influence on the development of modern ethology – in fact on all modern biology. Most people know him as the father of the theory of evolution, which in itself is the foundation for any study of animal behaviour. However, he also approached the subject more directly, and his last published work in 1872, *The Expression of the Emotions in Man and Animals*, was probably the first modern work on comparative ethology.

The Schools of the 20th Century

At the beginning of the 20th century, behavioural research grew quickly. However, development in the USA and Europe took different directions. American researchers were influenced by the behaviouristic approach, developed by people such as John B. Watson, and later Burrhus Frederic Skinner. Their work focused primarily on controlled experiments in laboratory environments, and their subject species *par preference* were rats and mice. Their interest centred on the mechanisms of learning and acquisition of behaviour through reinforcement or punishment (Goodenough *et al.*, 1993). Behaviouristic research was concerned with finding general rules and principles of learning, and there was a strong belief that such rules were independent of context. Therefore, the evolutionary history of the study subjects, or their ecological ways of life, were regarded as irrelevant for the research.

In contrast, the development of the science in Europe was dominated by naturalistic biologists, who spent most of their time observing wild animals in nature. Birds and insects were favourite subjects, and these researchers were mostly interested in instinctive, innate and adaptive behaviour. One of the pioneers was Oskar Heinroth, who first started to use the term 'ethology' with the meaning we give it today (Thorpe, 1979). The naturalistic behavioural biologists shared an important approach with the behaviourist. They were not particularly interested in the mental processes or emotions which may be associated with behaviour. Such processes were often regarded as unavailable for scientific research, since they were not considered to be observable. Only much later has a scientific interest for mental processes emerged, something which will be dealt with further in Chapter 4.

In the footsteps of Heinroth we meet two scientists whose influence over modern ethology cannot be overemphasized: Niko Tinbergen in Holland and Britain, and Konrad Lorenz in Austria. Tinbergen developed a field methodology of high precision. He designed experiments where details of the environments of free living animals were altered and their subsequent behaviour recorded. He was a pioneer in experimental ethology (Dawkins *et al.*, 1991). Lorenz, on the other hand, did not go much into nature with his research, but rather bred his experimental animals himself and kept many of them almost as pets. He rarely conducted elaborate experiments and was not prone to quantitative recordings. The strength of Lorenz was on the theoretical level. He formulated many of the fundamental ideas in ethology, and developed the first coherent theory of instinct and innate behaviour (Goodenough *et al.*, 1993).

Lorenz and Tinbergen definitely placed ethology on the solid ground of well-accepted sciences when, in 1973, they, together with the German researcher Karl von Frisch, were awarded the Nobel prize in medicine and physiology.

Modern Approaches to Ethology

From the 1960s onwards, ethology developed into the science it is today. This was guided to a large extent by the research programme formulated by Tinbergen, still generally accepted as a study of the basics of ethology (Tinbergen, 1963; Dawkins *et al.*, 1991). This programme is frequently referred to as 'Tinbergen's four questions', and the four aspects of behaviour that he felt were most important to ethology were:

1. What is the causation of the behaviour? The answer to this question refers to the immediate causes, such as which stimuli elicit or stimulate a behaviour, or which physiological variables, such as hormones, are important in the causation.
2. What is the function of the behaviour? In this context, the answer describes how the behaviour adds to the reproductive success, the fitness, of the animal. It therefore has to do with evolutionary aspects and consequences.
3. How does the behaviour develop during ontogeny? Studies of this question aim at describing the way a behaviour is modified by individual experience.
4. How does the behaviour develop during phylogeny? This is clearly an evolutionary question, and usually calls for comparative studies of related species.

Whereas early ethology was concerned mainly with causation, ontogeny and phylogeny, research during the 1960s and onwards became more and more focused on the functional question. Researchers have outlined new theories on how behaviour evolves through individual selection on the gene level, and have provided formal mathematical models for how the functional aspects of behaviour could be determined. The impact of this approach on contemporary animal behaviour science has been tremendous.

One aspect that was not covered by Tinbergen's questions was what animals perceive, feel and know in relation to their own behaviour. As mentioned earlier, this aspect of animal behaviour was largely considered to be unaccessible for science. However, other scientists have developed methods and concepts to allow investigation into this area. This has led to a new branch of the science, emerging in the 1970s, known as cognitive ethology (Bekoff, 2000) (the word cognition means subjective, mental processes – or thinking).

Applied Ethology

Early in the development of ethology, it was already apparent that the new insights into the biology of behaviour could be of great value in understanding more of the behaviour of domestic animals. This branch of science went into a dramatic expansion as the debate on animal welfare in so-called factory farming (a concept coined by the influential writer Ruth Harrison) developed in the 1960s. However, applied ethology is not only concerned with animal welfare. Let us look at a few areas of interest.

Welfare assessment

There is no doubt that the welfare of animals on farms, in zoos and laboratories, dominates the interest of most researchers in the area. The problems may be formulated, for example, like this: most laying hens in the world are kept in cages made of wire mesh, with very little space available for the animals and almost no substrates for carrying out many of the species-typical behaviour patterns of poultry (Fig. 1.2). So what are the most essential behaviour patterns for laying hens? Perhaps it is being able to dust bathe, or to perch during the night, or to perform nest building and laying the eggs in a secluded area. All of these are typical poultry behaviours, and there may also be others. How are the animals affected if they cannot behave like this? Can the activities be rated in any order of importance to the animals (Appleby *et al.*, 1993)?

Furthermore, a common alternative to battery cages is a floor housing system, with thousands of hens in one big group, sometimes with quite high stocking rates. In this situation, some unwanted behaviour (which can be present both in cages and in floor systems) may cause great harm to the animals, such as feather pecking or cannibalism. So, is it better for the hens to be in a situation where they can perform all the activities mentioned above, but where the social system may collapse and abnormal behaviour may spread widely (Hansen, 1994)?

Difficult questions such as these are important aspects of welfare assessment – only rarely do all indices point in the same direction. Chapter 6 will examine these aspects further, and describe some of the methods that researchers have developed to try to answer such questions.

Fig. 1.2. Battery cages and floor housing systems both cause behavioural problems for laying hens. To estimate the relative importance of different behaviours to animals, thereby allowing better decisions regarding choice of housing systems, is one important goal of many researchers in applied ethology.

Optimizing production

Farm animals are kept to produce food and other essentials for humans, and the farmers need to make profit from their enterprises. It is therefore necessary that the difference between the value of what the animals produce (for example, amount of milk or meat) and the costs that the farmer incurs for this production (for example, feed, investments and labour) is sufficiently high.

By taking animal behaviour into account, such optimization may be easier to achieve. For example, animals may utilize their feed better if they are fed according to their species-specific feeding rhythm and in a social context to which the species is adapted (Nielsen *et al.*, 1996). Social animals may eat more and digest the food better when all in a group are allowed to feed simultaneously.

Social animals that are kept in individual housing systems may be poorer at transforming feed into valuable products. Likewise, husbandry routines applied at a biologically inadequate time may decrease the production rate of the animals. Young piglets that are weaned from their mothers too early and in an abrupt manner show a decreased growth curve, and mixing of piglets after weaning may also have negative results on production (Algers *et al.*, 1990; Pajor *et al.*, 1991).

Behavioural control

The essence of keeping animals in captivity is to control their behaviour – by preventing their escape, controlling their breeding and making them adapt to the housing environment. The control is achieved largely by direct human actions, but also by the use of technical equipment.

A growing interest has been paid to the nature of human–animal interactions. For example, researchers have investigated how animals perceive humans and how they remember experiences with human behaviour. This may help farmers and others to interact more smoothly with their animals (Hemsworth and Barnett, 1987).

An increasing trend in animal farming is to use technical inventions in taking care of the animals. For example, group-housed pregnant sows are often fed from an electronic feeding station, which, to some extent, the animals are required to control themselves. The sows are equipped with transponders that allow them to open the feeding stations and obtain their individual feed rations. However, such systems must be designed carefully to avoid problems. For example, the social hierarchy of a group of sows may lead to some sows occupying the feed entrance, biting and wounding other animals and thereby destroying the functionality of the system. By using ethological knowledge, technical equipment can be designed to work better for the animals (Broom *et al.*, 1995).

Behavioural disorders

Housing systems such as those described earlier, or malfunctioning technical equipment or poor human management may all lead to various behavioural disorders. Aggression levels may become excessively high and dramatic behaviour such as cannibalism and several other types of abnormal behaviour may develop (Lawrence and Rushen, 1993). This is not only the case for farm animals. Many pets develop unwanted and abnormal behaviour, such as owner-directed aggression, uncontrolled urination and defecation in the home, or anxiety-like states.

The characterization and understanding of abnormal behaviour is a central aspect of applied ethology. Sometimes the behaviour can be cured by behavioural therapies, such as enriching of the home environment or stimulation of other behaviour. At other times, the research can provide insights that may help in avoiding the development of the behaviour.

The Field of Applied Ethology

As should be clear from these accounts, ethology is a science that may offer many different sorts of applications in situations where humans utilize animals for various purposes. Whereas animal welfare assessment clearly dominates in the applications of this science, it is by no means the only way in which knowledge of behaviour can be used.

Applied ethologists are normally concerned with all four of Tinbergen's questions. The causation and ontogeny of behaviour are essential aspects of understanding, for example, how abnormal behaviours develop and how they can be prevented. Phylogeny and function of behaviour are often less emphasized, but many studies have advanced our understanding of domestic animal behaviour greatly by considering how it can have evolved in the ancestors of these animals, and how it may have been affected by domestication (Fraser *et al.*, 1995). Experimental studies tend to dominate, but important scientific data have been made available through studies of domestic animals in wild-like conditions (which will become obvious in Chapters 7–13, where accounts are given of the normal behaviour of some important domestic species).

In the optimal situation, applied ethology research concerns every one of the fields outlined above. They are all inter-linked: poor human–animal, or equipment–animal, interaction may cause poor welfare, which in turn leads to behavioural disorders and reduced production. Applied ethology is therefore an essential part of the proper keeping of animals; and, last but not least, as will become obvious in this book, understanding the behaviour of domestic animals is a fascinating aspect of biology in its own right.

References

Algers, B., Jensen, P. and Steinwall, L. (1990) Behaviour and weight changes at weaning and regrouping of pigs in relation to teat quality. *Applied Animal Behaviour Science* 26, 143–155.

Appleby, M.C., Smith, S.F. and Hughes, B.O. (1993) Nesting, dust bathing and perching by laying hens in cages: effects of design on behaviour and welfare. *British Poultry Science* 34, 835–847.

Bekoff, M. (2000) Animal emotions: exploring passionate natures. *BioScience* 50, 861–870.

Broom, D.M., Mendl, M.T. and Zanella, A.J. (1995) A comparison of the welfare of sows in different housing conditions. *Animal Science* 61, 369–385.

Dawkins, M.S., Halliday, T.R. and Dawkins, R. (eds) (1991) *The Tinbergen Legacy.* Chapman & Hall, London.

Fraser, D., Kramer, D.L., Pajor, E.A. and Weary, D.M. (1995) Conflict and cooperation: sociobiological principles and the behaviour of pigs. *Applied Animal Behaviour Science* 44, 139–157.

Goodenough, J., McGuire, B. and Wallace, R.A. (1993) *Perspectives on Animal Behavior.* John Wiley & Sons, New York.

Hansen, I. (1994) Behavioural expression of laying hens in aviaries and cages: frequency, time budgets and facility utilisation. *British Poultry Science* 35, 491–508.

Hemsworth, P.H. and Barnett, J.L. (1987) Human–animal interactions. In: Price, E.O. (ed.) *Veterinary Clinics of North America: Food Animal Practice*, Vol. 3. W.B. Saunders, Philadelphia, pp. 339–356.

Lawrence, A.B. and Rushen, J. (eds) (1993) *Stereotypic Animal Behaviour – Fundamentals and Applications to Welfare*. CAB International, Wallingford, UK.

Nielsen, B.L., Lawrence, A.B. and Whittemore, C.T. (1996) Feeding behaviour of growing pigs using single or multi-space feeders. *Applied Animal Behaviour Science* 47, 235–246.

Pajor, E.A., Fraser, D. and Kramer, D.L. (1991) Consumption of solid food by suckling pigs: individual variation and relation to weight gain. *Applied Animal Behaviour Science* 32, 139–156.

Thorpe, W.H. (1979) *The Origins and Rise of Ethology*. Heinemann Educational Books, London.

Tinbergen, N. (1963) On aims and methods of ethology. *Zeitschrift für Tierpsychologie* 20, 410–433.

2 Behavioural Genetics, Evolution and Domestication

Per Jensen

As we noted in Chapter 1, people have been amazed for hundreds of years over the fact that animals are often able to perform extensive and complex, seemingly goal-directed, behaviour with no prior learning possibilities. Birds may build elaborate nests and migrate to traditional wintering sites even if they are raised out of contact with other members of the same species. Darwin suggested that such 'instinctive' behaviour was somehow transferred from the parents without the need for any learning processes, and hence, 'instincts' would be suitable raw material for evolution. Behaviour would then evolve as an adaptive trait, just like any morphological or physiological character.

When Darwin formulated his ideas, he knew nothing about genes. The work of the 'father of genetics', Gregor Mendel, did not become known to the scientific world until several decades after Darwin's death, and the discovery of the chemical structure of DNA was still almost 100 years away. However, Darwin's suggestion was prophetic: we now know that genes, composed of DNA, contain the codes for behaviour, and that evolution modifies the frequency of genes over generations, and therefore moulds the behaviour of species and individuals. In this chapter, we will examine some of the evidence for this, and its implications.

The focus of this book is on domesticated animals, so of course we will need to ask how domestication has come about, and how animals have been changed as a consequence of being domesticated.

To What Degree is Behaviour Genetically Inherited?

Anyone interested in dogs and dog breeds would have no problems in accepting the idea that even complex behaviour is genetically inherited from the parents. Retriever puppies are, generally, more inclined to retrieve and carry things around than other breeds, while border collie puppies have a strong tendency to herd. Some breeds are more famous for being aggressive, while others are generally viewed as friendly or docile. Observations such as these are suggestive, but they could perhaps also be explained by, for example, the offspring learning a specific behaviour from their parents, or by the fact that some types of owners tend to keep some types of dogs, and therefore may affect their behaviour differently. So how can we tell whether, and to what extent, a behaviour is inherited genetically?

One of the more famous examples of a clear genetic influence on behaviour comes from studies of lovebirds, *Agapornis* spp. (Dilger, 1962). One species, Fischer's lovebird, carries nest material (for example, strips of paper) one piece at a time in the beak. The closely related peach-faced lovebird tucks the strips in between the rump feathers and is therefore able to carry more nest material at each flight. Hybrids between the two show a poorly functioning mixture of the behaviour. They attempt to tuck material between the feathers, fail to let go of it, pull it out again, and then start the sequence over again. After several months of practising, the behaviour can become at least partly successful, in that the parrots manage to transport some material back to the nest site, but not in a manner typical of either of the parents. The strange behaviour of the hybrids is consistent with an intermediate (non-dominant) inheritance pattern of one or more alleles (an allele is one variant of a gene that is present in the population; hence every gene may have several alleles).

Another source of evidence is the bulk of experiments where the tendency to behave in a certain manner has been artificially selected for over generations. Mostly, this leads to individuals of the selected lines being more and more inclined to behave according to the selection criteria, indicating a strong genetic basis for the behaviour. For example, in fruit flies (*Drosophila melanogaster*) mating time may vary considerably between individuals. In a classical experiment, 100 fruit flies of both sexes were placed in a chamber and the 10 fastest and the 10 slowest maters were determined. They were allowed to breed separately, and the procedure was repeated in each new generation. After 25 generations, mating time differed between populations by about 30 minutes (Fig. 2.1) (Manning, 1961). Similarly, the tendency for positive versus negative geotaxis, i.e. to move towards or against gravitation on a vertical surface, varies between fruit flies, and selection for the strength of this tendency produced rapid changes in the population over a few generations (Hirsch, 1967).

Experiments such as these indicate that even complex behavioural responses can be controlled by genes. Since there is a quantitative response – behaviour becomes gradually faster or slower, or a tendency is gradually strengthened or weakened – it is, in these cases, most likely that several genes interact in a quantitative fashion in producing the observed behaviour phenotype. Later, we will examine some cases where few, or even single, genes affect the behavioural output.

Do studies of insects have any relevance for applied ethology, which is mostly concerned with mammals and birds? Yes, genetic control is universal among animals. Not only are the tools – DNA, RNA, etc. – identical,

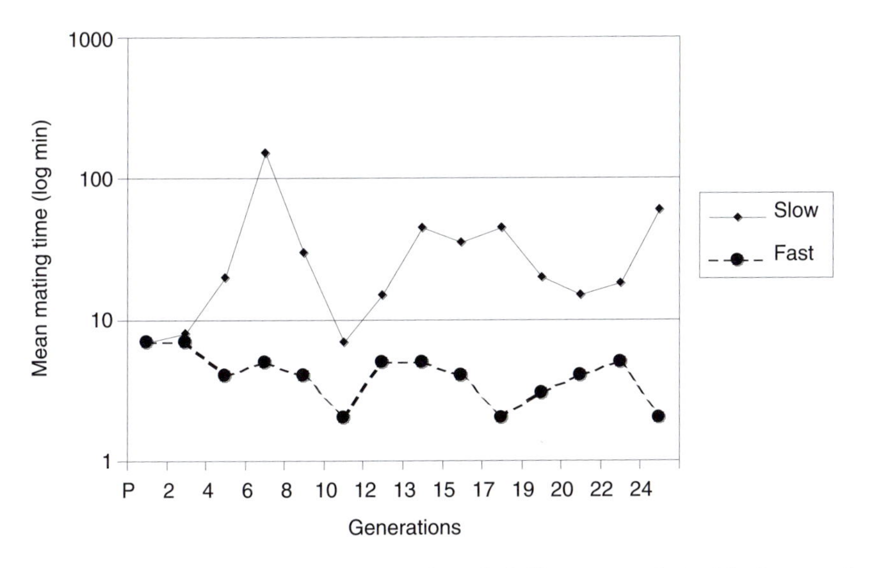

Fig. 2.1. Results of a selection experiment, where fruit flies were selected for long or short mating time (modified after Manning,1961).

the way in which the control is exerted is largely the same (we will deal with the mechanisms later in this chapter). Moreover, the sequencing of the complete, or almost complete, genomes of organisms, ranging from yeast, via the nematode *Caenorhabditis elegans* and fruit flies to humans, has revealed large similarities in structures and functions of many genes. Hence, genetic instructions for behaviour may sometimes be very similar across species and phyla.

There is also a lot of direct evidence of genetically determined behavioural differences among domestic animals. For example, laboratory mice were selected for their tendency to build nests. After 15 generations, one line collected about 50 g of cotton used for nest material, compared to 5–10 g in the other line (Lynch, 1980). As another example, laying hens sometimes develop a problematic behaviour called feather pecking, where they damage the feathers of flock mates (see Chapter 6). The tendency to develop the behaviour differs between strains, and genetic selection against the behaviour can cause it to decrease over a few generations (Kjaer and Sorensen, 1997). This indicates a genetic basis for this abnormal behaviour.

Genetic selection of desired behaviour, or against undesired behaviour, is therefore also an important means for improving animal welfare. For example, the level of fearfulness has been reduced by selection in both poultry and quail (Jones and Hocking, 1999). Fearfulness has also been reduced in some animals kept for fur production, such as mink, along with undesirable behaviours such as chewing and destroying the fur (Malmkvist and Hansens, 2001).

How can there be genes controlling the onset of abnormal behaviour such as feather pecking? Probably this is not how it works. We must always remember that even though selection is performed on a well-defined trait, the observed responses can often be explained by indirect effects. The reduction of feather pecking by selection has been suggested to be caused indirectly, by a general decrease in foraging activity which reduces the overall tendency of the birds to peck (Klein *et al.*, 2000). Therefore, in this case, we cannot be certain that there is direct genetic control of feather-pecking as such – in fact it may seem more logical to assume that the effect is actually indirect.

Genetic versus Environmental Influence on Behaviour

Sometimes, people misinterpret the fact that genes contain instructions for the behaviour of an animal. The misinterpretation is often referred to as genetic determinism, the belief that if there is a genetic control mechanism, the behaviour of an individual will be inflexible and determined from the point of fertilization. The following example illustrates nicely why this is wrong.

Rats will normally learn easily to run through a maze, in order to reach a goal box containing some food. With increasing number of trials, they will make fewer and fewer errors, in the sense that they become less and less likely to run into blind alleys. However, there is considerable individual variation in how easily this task is accomplished. In a famous experiment, rats were selected depending on how fast they learned a specific maze. In only a few generations, the selected lines had separated, with almost no overlap, into a line of so-called 'bright' rats and another of so-called 'dull' rats (Tryon, 1940). These lines could be preserved over generations, and the offspring would show the same pattern in learning ability as their parents.

A possible interpretation of this experiment is that learning capacity is controlled by genetic factors. Once it is known which population an animal descends from, whether its father and mother are dull or bright, it would be possible to predict accurately how well this animal would manage a maze-learning test. But this conclusion is oversimplified, as shown by the following later study.

Rats from the same populations were raised in three different environments: the standard laboratory environment in which the populations had been maintained for generations, an impoverished environment without bedding material or any other interesting stimuli, and an enriched environment, where different substrates for manipulation and stimulation were added to the cage (Cooper and Zubek, 1958). When these animals were tested in the maze, those reared in the standard environment showed the same differences between lines as expected: the 'dull' line had considerably larger difficulties with mastering the problem.

However, when the behaviour of the rats from the two lines reared in a restricted environment was compared, an interesting result emerged. The differences between the strains had disappeared, and they both performed as poorly as the dull rats reared in the standard environment. Furthermore, there were also no differences between the strains when they had been reared in enriched environments, but this time they both performed as well as the bright strain from standard cages (Fig. 2.2).

These results show how careful we must be in inferring deterministic genetic control over behaviour when a genetic correlation has been demonstrated. Clearly, genes supply the organism with the necessary basis for a particular behaviour – the limbs, muscles, nerves and sensory organs, and a central nervous system – but any behaviour is likely to develop in synchrony with the environment in which the animal lives. Rather than considering genes as determinants of behaviour, we should consider genetic traits to be predispositions, which bias animals towards certain reactions and developmental pathways.

This approach also helps in understanding the possible indirect genetic influence on feather-pecking, described earlier. If there is a genetic predisposition to peck at different substrates, feather-pecking may develop when the environment lacks necessary pecking stimuli. Hens in a stimulus-rich environment may therefore never develop the behavioural disturbance, although the same individuals would do so under poorer conditions.

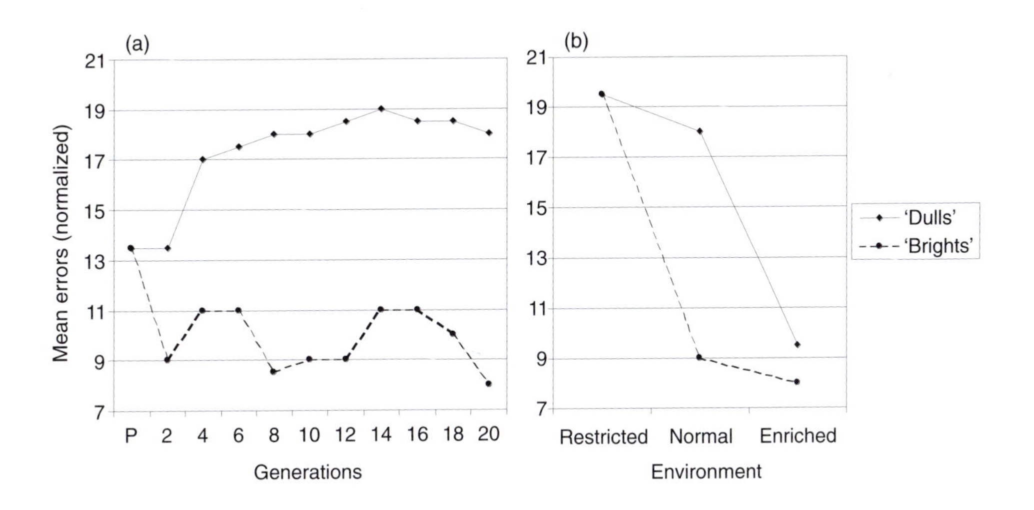

Fig. 2.2. (a) Results of selection for maze-learning capacity in rats (modified after Tryon, 1940). (b) Results of a maze-learning experiment where individuals from each of the selected lines were raised in either the standard laboratory environment, an impoverished laboratory environment or an enriched environment (modified after Cooper and Zubek, 1958).

Single Gene Influences

Since behaviours are complex traits, involving identification of stimuli acting on different sensory channels, central nervous processing of information and the concerted actions of groups of muscles, it may seem implausible that single genes would be able to control the expression of behaviour. Nevertheless, there are several examples of such effects of single, or at least of very few, genes.

In a comprehensive study of behavioural genetics, two breeds of dogs with very divergent behaviour, the basenji and the cocker spaniel, were crossed (Scott and Fuller, 1965). By producing back-crosses and using standard genetic analysis methods, it was possible to suggest simple genetic control mechanisms for rather complex behaviour patterns. For example, struggling when being restrained – more pronounced in basenjis – was consistent with inheritance by one single allele, showing no dominance. The tendency to bark was suggested to be inherited through two dominant alleles for a low stimulus threshold (basenjis are famous for having a very high barking threshold, i.e. they rarely bark).

In fruit flies there are many examples of effects of separate loci on complex behaviour. One mutant strain, called 'dunce', has a mutation in a single gene on the X chromosome. These flies cannot learn to avoid an odour when trained in an experiment that allows normal flies to connect the odour with an electric shock (Quinn *et al.*, 1974).

How do Genes Affect Behaviour?

Studies such as those summarized above demonstrate clearly that genes have a strong influence over behaviour. But what does a gene actually do to affect behaviour, trapped as it is inside the nucleus of the cells? How is it possible that mutations in one or a few genes can change complex patterns of behaviour? DNA is a fantastic molecule for storing information, but the information in a gene is only used for one thing: as a code for synthesis of proteins. What is the link between proteins and behaviour?

Since genes carry the codes for synthesis of proteins, they thereby regulate important aspects of metabolism, hormone concentrations, etc. Of course, the direct interaction between the environment and the animal, including its behaviour, mostly occurs via the nervous system. However, the wiring up of neurons into an entire nervous system, as well as the design and sensitivity of receptors and sense organs, are all processes that are regulated by the genes during embryology and further ontogeny. In the adult animal, levels of hormones, neurotransmitters and other substances which are extremely important for behavioural control are regulated by the genes.

The detailed pathway from the synthesis of a specific protein to an observed behaviour is long and complex, and we have actually only started to understand some of it. For example, the 'dunce' fruit flies, discussed above, have a defective gene that is normally responsible for the production or activity of an enzyme called cyclic AMP phosphodiesterase, which degrades cyclic adenosine monophosphate (AMP) (a common intermediate molecule in cell metabolism). So either the enzyme or the cyclic AMP itself is involved in olfactory learning. Once pathways such as these are more fully understood, they will no doubt provide deep insights into the biology of behaviour.

Even if we have seen examples of single-gene effects on behaviour, a certain behaviour is usually not controlled by a single gene. However, a single gene can have a crucial role for the normal appearance of a certain behaviour. Single mutations may alter the behaviour drastically, for example by changing the development of important structures (Huntingford, 1984). One mutation in the nematode *Caenorhabditis elegans* changes the myosin filaments in parts of the body and produces uncoordinated movements. Another mutation affects the structure of chemoreceptors in the head and changes the receptivity of the worm to chemical stimuli. The slow-mating fruit flies, which we discussed earlier, were suggested to have an altered production of juvenile hormone.

Localizing the Genes

Even if it is clear that genes predispose animals to specific types of behaviour, the mechanisms whereby this happens are still largely unknown, particularly in vertebrates. Usually we do not even know exactly which genes are involved in controlling a behaviour, and we therefore cannot understand what these genes might do to exert their effects. With the rapid development of genetics, identifying the genes and their functions is likely to be an important area of research in the immediate future (Fig. 2.3).

A first step towards identifying genes tied to specific behavioural traits is to localize the position of the genes on the chromosomes. The position can sometimes be determined because a behaviour is inherited in close correlation with some other trait, for which the gene and its location are known – what geneticists refer to as linkage. It could, for example, be a colour pattern or some other observable trait. One can then conclude that the behaviourally linked gene must be located close to the gene for the other trait.

With modern gene technology, new possibilities for gene localization have emerged. One common method is to use quantitative trait loci (QTL) analysis. With this analysis, the genomes of animals are mapped with respect to the occurrence of specific DNA-markers, usually non-coding sequences of base pairs, which vary between individuals in specific chromosomal locations. The correlation between a particular phenotypic trait, for example a behaviour, and the occurrence of different markers can then be assessed (sounds simple, but requires many animals, a lot of lab magics and advanced statistical computing). When a certain trait has been found to be correlated with the occurrence of a certain marker, one can imply

Fig. 2.3. When jungle fowl are crossed with modern laying hens, and the F_1 generation is subsequently intercrossed, the resulting F_2 generation exhibits much genetic and phenotypic variation. Such breeding experiments are important tools in QTL studies, i.e. experiments designed to localize the chromosomal loci of genes controlling specific traits.

that the gene that affects the trait is located close to that marker. By using the increasingly detailed gene maps which are available for more and more species (humans, for example), relevant genes in that chromosome region can be identified, and their products studied.

QTL analyses have identified chromosomal regions influencing many aspects of behaviour, such as the tendency of honey bees to sting, the cyclicity of activity in mice, preference for alcohol in mice and the hyperactivity syndrome of rats.

Methods of identifying the actual genes and their products are developing rapidly, and in the next few years we will most likely see a tremendous increase in our understanding of behavioural genetics. One important source is going to be so-called knock-out animals. These are strains of animals where specific genes have been made silent. The development of such an animal, compared to normal ones, therefore tells us important things about the function of the knocked-out genes. As an example, mice lacking the gene for oxytocin (a hormone involved in many different processes such as parturition and milk ejection) not only – as expected – lack the capacity to eject milk, but also show a reduction in aggressive behaviour (Crawley, 1999).

Evolution of Behaviour

Now that we have seen how behaviour depends on genetic predispositions, we have two of the necessary keys to understand how behaviour can have developed during evolution. It is nowadays basic biological knowledge that animals are products of a long evolutionary history, whereby their anatomy and physiology have been adaptively shaped to what we see today. But is that also the case for behaviour?

In order for any trait to be modified by evolution, three principles are required, which can be deducted from Darwin's original writings:

1. The principle of variation. This states that a trait must vary between the individuals of a population. If all individuals are identical, no evolution of the trait is possible.
2. The principle of genetic inheritance. This principle requires that some of the variation in the population must be of genetic origin. It is not necessary that the trait is genetically determined, only that genes have some influence over the phenotypic expression of the trait. It follows from this principle that, on average, when a trait is genetically inherited, individuals resemble their parents more than they resemble other randomly chosen individuals of the population with respect to this trait. Furthermore, the closer the relationship between any two individuals, the larger the resemblance.
3. The principle of natural selection. According to this, some variants of the trait must cause variation in the ability of individuals to reproduce. If the reproduction capacity is enhanced, the trait will increase in frequency over generations, and if it is reduced, the frequency will decrease.

In order for evolution to modify any trait, all these principles must be fulfilled simultaneously. In fact, when they are all fulfilled, evolution is bound to happen. In the case of behaviour, we have already seen that there is often a large variation within populations (for example, in the case of maze-running ability in rats), and that this variation is often partly caused by genetic differences between individuals. What remains is to show that this genetically based influence causes different reproductive success.

Indeed, most of contemporary evolutionary research in behaviour is devoted to examining the reproductive advantages of having a certain behaviour rather than any other possible alternative. Obtaining and defending a territory may increase the chance of getting a mate and reproducing, and territory quality and size is often found to be closely related to reproductive success. Searching for food in a manner that causes minimal exposure to predators may seem an obvious example, and different food searching patterns do have effects on the efficiency of reproduction. Some decades of intense research have produced innumerable similar examples of behaviour patterns which follow the principle of natural selection, and this is often covered in the subject called behavioural ecology (Krebs and Davies, 1991).

Tracing the Evolutionary History of Behaviour

Behavioural archaeology is a rather fruitless field of science, but occasionally fossils may actually carry some information about the behaviour of long-extinct animals. For example, close examinations of the skeleton and feather structure of the oldest bird fossil, *Archaeopteryx*, has led to the suggestion that bird flight developed through gliding from tree branches, and running and leaping after prey on the ground (Alcock, 2001) (although other interpretations have also been suggested). Mostly, however, we have to rely on comparisons with animals that are living and behaving around us today.

The method applied in these kinds of studies is to take advantage of phylogenetic trees, which may be deduced from traits other than behavioural, and then map the behaviour seen in closely related species on to this tree. For example, in one study of courtship displays in manakins, a group of small, fruit-eating tropical birds, a phylogenetic tree was constructed based on similiarities between species in the structure of the vocal apparatus (syrinx). The male courtship behaviour of the different species was then examined in detail. There were 44 behavioural elements that occurred in at least one of the 28 examined species. By mapping the species to the phylogenetic tree based on how many of the behavioural traits they shared, it was possible to deduce a possible history of how present-day courtship has evolved from simpler behaviour in different ancestors. Mostly, evolution has led to more elements being included in the repertoire, but in some cases, behavioural patterns have been lost during evolutionary development (Prum, 1990).

Sometimes, the mechanisms forming behavioural differences between species have been possible to elucidate by comparative studies. The greenish warbler, *Phylloscopus trochiloides*, occurs in several subspecies throughout Russia, Siberia and China. In its western range, the subspecies interbreed readily. However, during evolutionary time, subpopulations have been separated by the Tibetan Plateau, and where the populations meet again, they are in effect two different species. They produce completely different songs, and the females do not recognize the songs of the other population as species-specific (Irwin *et al.*, 2001). Geographical separation and female partner choice has driven the formation of a new warbler species.

Modification and Ritualization

The reconstruction of the evolutionary history of behavioural traits leaves us with some important lessons. First, evolution works only by selection of variants of traits which are already present in an animal. This means that evolution cannot invent any new behaviour in the face of a 'need'. Hence, the reason why an animal uses a particular behaviour for a certain purpose is often to be found in the history of the species.

Inevitably, all behaviours we see today are therefore modified versions of behaviour patterns that may have served very different purposes in ancestral forms. A behaviour that serves a certain function becomes slightly modified and may then serve a slightly different function and, as time and generations pass, a new behaviour has developed. We can verify this causative chain since the original behaviour may be conserved in closely related species, with its original function still intact.

A special case of this is the evolution of animal signals by the process of ritualization. One of the most quoted examples of a ritualized display is that of the peacock's courtship display (Fig. 2.4). The enormous and colourful tail is spread like a fan, vibrating, at the same time as the cock is bowing towards the female. A likely evolutionary history looks like this: ancestors attracted females by pecking at feed items on the ground and emitting a special call – this is still the common courtship of one close relative, domestic poultry. These movements have become exaggerated, as in pheasants, and finally become unrelated to feed presentation, as in the peacock. The tail has developed and increased in size at the same time. A behaviour that originally served as food presentation has now become the courtship of peacocks (Schenkel, 1956). Ritualization is therefore the process by which a certain behaviour evolves into a signal by becoming exaggerated and losing its original function.

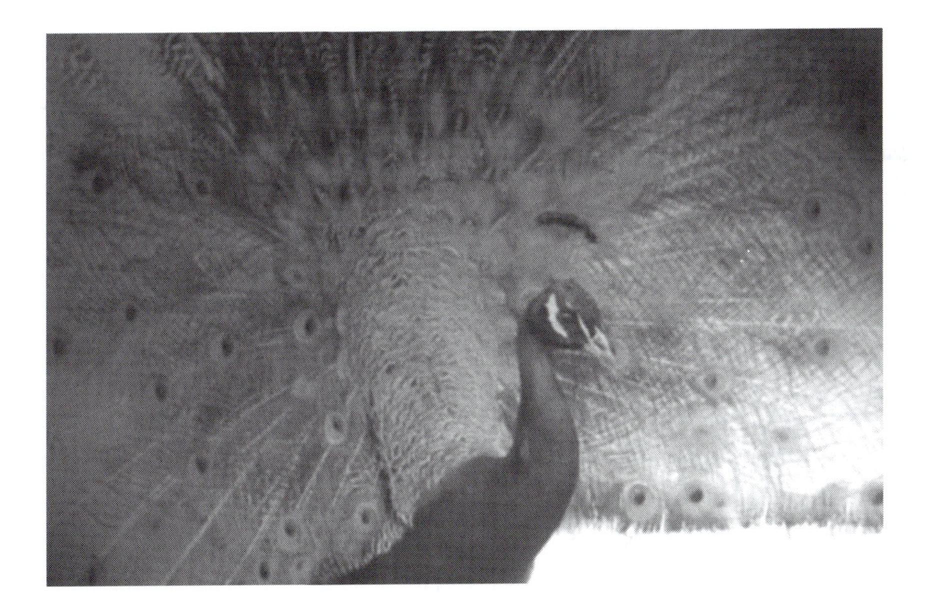

Fig. 2.4. Domestic cockerels court females in the same way as jungle fowl. One important element is 'tid-bitting', where the male emits a special call at the same time as he pecks towards the ground, sometimes towards food particles. This has, by ritualization, developed into the spectacular peacock courtship.

Ritualization, combined with exaggeration of anatomical traits (such as the male peacock tail), is also driven by preferences of the females, which, since Darwin, is referred to as sexual selection. A trait, for example a colour or an ornament, may evolve because they indicate important qualities in the males. Perhaps only males that are genetically healthy and resistant to parasites can grow large tail feathers – the female that prefers to mate with males carrying large tails may then have offspring that carry the favourable genes (Alcock, 2001).

The Function of Behaviour

Another important lesson we have learnt from evolutionary biology is that behaviour does not evolve for 'the good of the species'. Since all evolution can do is to work on variation between individuals, there is simply no mechanism around that can work on behalf of a group of animals or a species (but there are noteworthy exceptions, where the group may be an efficient unit of selection; see Chapter 5 for more details). When we look for the function of any particular behaviour, we therefore always have to consider the benefits to the individual. That benefit ultimately has to be measured in reproduction. The reproductive success of an individual is also referred to as its fitness.

A behaviour not only has potential fitness benefits to the performer, but also costs. It may consume valuable energy, which could otherwise be used for reproduction, or expose the animal to predators. Once we realize that a behaviour has both costs and benefits, it becomes obvious that evolution will select the behaviour that maximizes the difference between fitness benefits and costs. This is called the optimal behaviour.

For example, big territories are usually better, since they provide more food and other resources. Consequently, females often prefer males that hold large territories. Should a male therefore always attempt to defend as big a territory as possible? No, big territories require much more energy and time to defend, leaving less time for reproductive efforts. Therefore, males should aim for optimal territories, which maximize the benefits received from attracting more females in relation to the costs (Fig. 2.5). Here is another example: how long should an animal continue to search for food in the same place rather than moving to another, perhaps more profitable site? The answer is that the animal should weigh up the information available regarding how depleted the present food patch is (number of food items ingested per time unit), how profitable another average patch is likely to be in the habitat, and how long it will take to find another patch. Then it should choose to leave the present patch when net energy intake over time is maximized. This general idea is central to what is called optimal foraging theory (Krebs and Davies, 1991).

It may seem implausible that animals would be able to perform complicated calculations like this. Nevertheless, many studies have shown

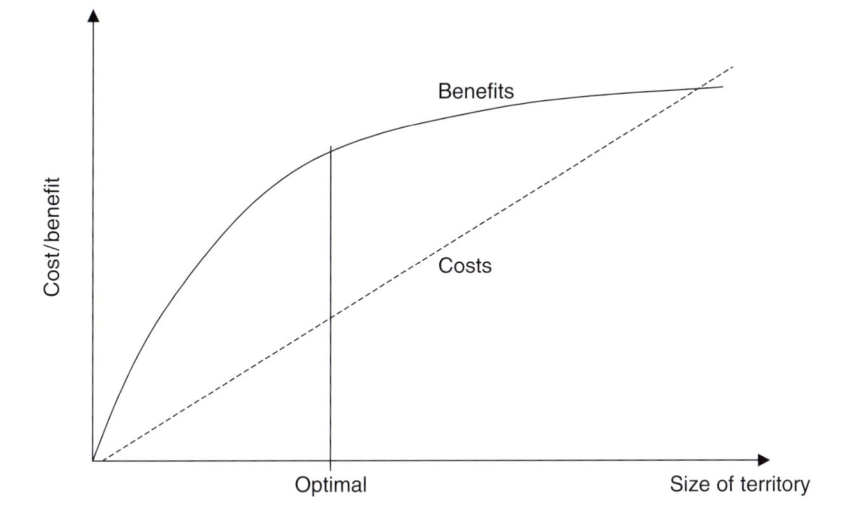

Fig. 2.5. Costs and benefits of a territory both increase as a function of the size of the territory, as is shown in this principle diagram. The functions are not identical – cost often increases linearly or with an increasing function, whereas benefits follow a decreasing function. This is because an animal cannot gain infinitely from larger territories. There are, for example, time constraints. Evolution has favoured animals which choose to defend territories with an optimal size, i.e. where the difference between benefits and costs is largest.

that animals generally behave in an optimal manner (Krebs and Davies, 1991). But of course, they do not perform the mathematics implied in the theory. Evolution has selected animals with the highest fitness, and they are the optimal ones. If we could ask the animal why it leaves a patch after a certain time, it would not be able to tell us the evolutionary reason. Could it speak, it would probably say that 'it felt like leaving now'. In the technical jargon of behavioural ecology, the animal has a behavioural strategy that causes it to react to stimuli as if it was consciously maximizing fitness by optimizing its behaviour.

The same reasoning can be used to understand social behaviour. What is best for an animal to do in a social context (for example, to attack or retreat), depends strongly on what the opponent does or is likely to do. Even though it may superficially appear beneficial to always attack and drive away intruders from a food source, some further thinking shows us that this is not necessarily so. Since the tendency to attack depends on genes, a population where all individuals always attack each other may be very unstable. A mutation that avoids fights may gain a lot in fitness because it is not wounded, and the mutated gene may therefore spread quickly in the population. The optimal social strategy is therefore not always easy to predict without the use of formal mathematical modelling. The strategy which, on average, confers the largest benefit to the individuals of the population is usually referred to as the evolutionarily stable strategy, or ESS (Krebs and Davies, 1991).

Domestication – an Evolutionary Case History

Domestication, the process whereby an animal is transformed from a life in the wild to a life under some control of humans, is one of the most dramatic evolutionary processes accessible to scientific investigation. It is often treated in the literature as a cultural phenomenon, and the story often goes something like this: when humans became sedentary and agriculture started, some 10,000–15,000 years ago, those animals that were available for exploitation were tamed by humans, so they could provide food, clothing, etc. This led to the first domesticated species – dogs, goats and sheep – later followed by the main agricultural species, such as pigs, cattle and chickens. The reason why the story is often told in this way is that bones with typical domesticated traits (shortened legs, compressed skulls, etc.) are found in excavations of early agricultural sites. According to this 'standard model' of domestication, each domestic species emerged from one single ancestral species and from one single domestication event, or from a few events. However, a biological approach to the subject reveals a somewhat different story.

Modern DNA technology has provided new tools for a examining a variety of questions. For example, differences in sequences of DNA in animal mitochondria (mDNA, which is inherited only from the mother) reveal their degree of relatedness and also the time that has passed since the animals became reproductively isolated from each other. Examinations of such data show that, for example, dogs and pigs dissociated from wolves and wild boars much earlier than archaeological evidence has suggested. More than 50,000 years ago, wolves that were later going to develop into dogs became reproductively isolated from the line that led to today's wild wolves. Domestication of pigs seems to have happened independently in Europe and Asia and, again, reproductive isolation of the populations occurred much earlier than indicated by archaeology (Giuffra *et al.*, 2000). Independent events of horse domestication have been suggested from similar sorts of data (see Chapter 8).

This is consistent with a theory of domestication claiming that at least some animals in effect 'chose to be domesticated' (Budyansky, 1992), in the following sense. Some populations seem to have been adapted to take advantage of a newly arising, highly productive yet ephemeral habitat. The new species, *Homo sapiens*, emerging on the arena of life between 100,000 and 200,000 years ago, may have provided a potentially rich and fruitful niche for such animals. They may have enjoyed large fitness benefits from associating with humans, simultaneously increasing human fitness by being relatively easy to hunt or by guarding against danger (in the case of wolves). It appears that these populations became reproductively isolated from the rest of the species long before domestication has been thought to begin. Perhaps domestication took a leap when humans became sedentary and agriculture started, and perhaps what archaeologists discover may be the results of the onset of active breeding. However, domestication as an evolutionary process seems to have been going on for perhaps five or ten times longer than this (Clutton-Brock, 1999).

Which Animals were Domesticated?

One feature of domestication that requires a biological explanation is the similarity in some specific behavioural and ecological traits of the animals. Only some animal species appear to be suitable for domestication. From a systematic perspective, there is a large bias among domesticated animals towards ungulate mammals and gallinaceous birds. From a behavioural aspect, there is a huge dominance of gregarious omnivores or herbivores without strong mating bonds.

These behavioural traits (which are abundant among ungulates and gallinaceous birds) may be crucial for successful domestication (Price, 1997). Social life allows many animals to live together in the human setting, and predisposes for hierarchical systems, where humans can more easily adopt the role of a dominant group leader. Feeding habits that do not compete with humans would be essential for successful cohabitation. Weak bonds between mates would make selective breeding much easier.

The first wave of domestication comprised the major present-day farm animals, and animals such as the dog and the horse. When agriculture was established in large parts of the world, many of these animals were already present, and from a biological perspective, could be considered domesticated. For thousands of years, few new species were added (but of course a variety of breeds developed within each species). During recent centuries, a second wave has occurred, and a number of new species were domesticated. This time, there is no doubt that this was a process controlled completely by humans, dictated by specific needs or wishes. The process gave us fur animals (mink, foxes, racoon dogs, chinchillas, etc.), laboratory animals (mainly mice and rats) and several new meat producers (buffaloes, ostriches and salmon). The animals domesticated in this second wave do not necessarily show the typical traits outlined earlier. For example, mink are solitary, territorial animals, and foxes have strong mate bonds.

Behavioural Effects of Domestication

Since domestication involves a genetic modification of a population of animals, we would expect domestic animals to differ in a number of traits from their wild ancestors (Fig. 2.6). Typical morphological changes in colour (more white and spotted individuals), shape (size and relative leg length), and function (less pronounced seasonality in reproduction) may lead us to expect that there will be pronounced behavioural differences as well. However, most research has found only subtle differences

Fig. 2.6. Domestic pigs differ in appearance from wild boars, but their behaviour is very similar.

between domestic and wild animals (Price, 1997). Most typically, these differences can be attributed to modified stimulus thresholds, causing some behaviour patterns to become more common and others to be more rare during domestication. No new behaviours seem to have been added to the behavioural repertoire of any domestic species, and few of the ancestral behaviour patterns, if any, have disappeared completely.

Hence, pigs kept for generations in restricted indoor housing systems still build elaborate farrowing nests if released into a forest, and laying hens kept in battery cages will attempt to perch high up during the night if given the slightest opportunity.

The threshold differences that have occurred are sometimes caused by active selection by humans. For example, most dogs have been bred to bark in response to very low stimulation, whereas basenjis, used for sneak-hunting in Africa, have been bred for the opposite. Other behavioural changes may be the result of an evolutionary adaptation. For example, energetically costly behavioural strategies appear, in some cases, to have been reduced in pigs and poultry (Gustafsson *et al.*, 1999; Schütz and Jensen, 2001).

Although there are behavioural differences between wild and domestic animals, it is clear that they are not as large as we sometimes tend to believe. It is also frequently suggested that domestic animals are less responsive to their environment than wild animals, even that they are more 'stupid'. However, detailed studies of the behaviour of domestic animals in natural conditions reveal that their behaviour is very similar to that of their ancestors. The fact that they may often develop abnormal behaviour and even pathologies when they are prevented from performing a normal behaviour, indicates a strong responsiveness towards the environment in which they are kept. Regardless of whether we think that welfare concerns are central in animal husbandry, or whether we are guided by concerns over the productivity of the animals, we need to remember that the behaviour of domestic animals is controlled by genetic mechanisms shaped over thousands and thousands of generations of evolution in the wild, and only slightly altered during domestication. The evolutionary history and adaptations of the ancestors, and the natural behaviour of the present-day animals, are therefore important pieces of information if we want to understand the animals that we keep for our use.

References

Alcock, J. (2001) *Animal Behaviour – an Evolutionary Approach.* Sinauer Associates Inc., Sunderland, Massachusetts.

Budyansky, S. (1992) *The Covenant of the Wild – Why Animals Chose Domestication.* William Morrow and Co., New York.

Clutton-Brock, J. (1999) *A Natural History of Domesticated Mammals.* Cambridge University Press, Cambridge.

Cooper, R.M. and Zubek, J.P. (1958) Effects of enriched and restricted early environments on the learning ability of bright and dull rats. *Canadian Journal of Psychology* 12, 159–164.

Crawley, J.N. (1999) Behavioral phenotyping of transgenic and knockout mice. In: Jones, B.C. and Mormède, P. (eds) *Neurobehavioral Genetics – Methods and Applications.* CRC Press, Washington, DC, pp. 105–119.

Dilger, W.C. (1962) The behavior of lovebirds. *Scientific American* 206, 88–98.

Giuffra, E., Kijas, J.M.H., Amarger, V., Carlborg, Ö., Jeon, J.-T. and Andersson, L. (2000) The origin of the domestic pig: independent domestication and subsequent introgression. *Genetics* 154, 1785–1791.

Gustafsson, M., Jensen, P., de Jonge, F.H. and Schuurman, T. (1999) Domestication effects on foraging strategies in pigs (*Sus scrofa*). *Applied Animal Behaviour Science* 62, 305–317.

Hirsch, J. (1967) *Behaviour Genetic Analysis.* McGraw Hill, New York.

Huntingford, F. (1984) *The Study of Animal Behaviour.* Chapman & Hall, London.

Irwin, D.E., Bensch, S. and Price, T.D. (2001) Speciation in a ring. *Nature* 409, 333–337.

Jones, R.B. and Hocking, P.M. (1999) Genetic selection for poultry behaviour: big bad wolf or friend in need? *Animal Welfare* 8, 343–359.

Kjaer, J.P. and Sorensen, P. (1997) Feather pecking behaviour in white leghorns, a genetic study. *British Poultry Science* 38, 333–341.

Klein, T., Zeltner, E. and Huber-Eicher, B. (2000) Are genetic differences in foraging behaviour of laying hen chicks paralleled by hybrid-specific differences in feather pecking? *Applied Animal Behaviour Science* 70, 143–155.

Krebs, J.R. and Davies, D.B. (1991) *Behavioural Ecology: an Evolutionary Approach.* Blackwell Scientific Publications, Oxford.

Lynch, C.B. (1980) Response to divergent selection for nesting behaviour in *Mus musculus. Genetics* 96, 757–765.

Malmkvist, J. and Hansen, S.W. (2001) The welfare of farmed mink (*Mustela vison*) in relation to behavioural selection: a review. *Animal Welfare* 10, 41–52.

Manning, A. (1961) The effects of artificial selection for mating speed in *Drosophila melanogaster. Animal Behaviour* 16, 108–113.

Price, E.O. (1997) Behavioural genetics and the process of animal domestication. In: Grandin, T. (ed.) *Genetics and the Behaviour of Domestic Animals.* Academic Press, London, pp. 31–65.

Prum, R.O. (1990) Phylogenetic analysis of the evolution of display behavior in the neotropical manakins (Aves: Pipridae). *Ethology* 84, 202–231.

Quinn, W.G., Harris, W.A. and Benzer, S. (1974) Conditioned behavior in *Drosophila melanogaster. Proceedings of the National Academy of Sciences USA* 71, 708–712.

Schenkel, R. (1956) Zur Deutung der Phasianidenbalz. *Ornithologische Beobachtungen* 53, 182.

Schütz, K. and Jensen, P. (2001) Effects of resource allocation on behavioural strategies: a comparison of red junglefowl (*Gallus gallus*) and two domesticated breeds of poultry. *Ethology* 107, 753–765.

Scott, J.P. and Fuller, J.L. (1965) *Genetics and the Social Behavior of the Dog.* University of Chicago Press, Chicago.

Tryon, R.C. (1940) Studies in individual differences in maze ability VII. The specific components of maze ability and a general theory of psychological components. *Journal of Comparative Physiology and Psychology* 30, 283–335.

3

Physiology, Motivation and the Organization of Behaviour

Frederick Toates

Introduction

Think of observing animal behaviour and consider the different activities in which animals engage: mating, feeding, sleeping and grooming, etc. What are the determinants of these different forms of behaviour? What kind of process underlies the switch from one behaviour to another? The notion of *motivation*, the central topic of the present chapter, is involved in answering such questions (McFarland and Sibly, 1975). Motivation is sometimes termed a 'hypothetical variable', in that it cannot be observed directly. However, it has proven a useful concept in theorizing about the causes of behaviour. The term 'motivation' is used to describe internal processes that arouse and direct behaviour. It refers to an internal process that underlies the tendency to engage in a particular behaviour, e.g. feeding 'motivation' is said to underlie feeding. Motivation is produced by an interaction of internal factors (e.g. low nutrient level) and external factors (e.g. food presentation, a cue that food will soon be available). Asking about the factors that produce motivation, and thereby a tendency to show a particular behaviour, are 'how' questions and form our primary concern. However, the notion of the function of behaviour ('why' questions) is also relevant to the explanation of motivation and behaviour, and so will be addressed briefly.

Motivation is of crucial importance to applied ethology since, amongst other reasons, it is well recognized that animals have motivations for behaviour beyond that which is strictly necessary for bodily survival. For example, theorists speak of such motivations as those to escape, explore and play. The mode of expressing motivation can also be important. In deriving satiety from feeding, the ability, say, to gain food by using natural species-typical behaviour might matter in addition to any nutrients that are gained.

Although motivation is the central topic of this chapter, of course such a process does not act in isolation in determining behaviour. Therefore, to contextualize the subject, this chapter includes consideration of various determinants of behaviour. That is to say, it concerns the nature of the different processes that trigger behaviour and how they interact. The chapter will suggest a classification of such determinants, with a focus upon the role of motivation.

Behaviour is the outcome of the activity of skeletal muscles that effect action on the external world. Thus, any controls of behaviour must ultimately act at the level of the associated motor neurons. With this in mind, another concern of the chapter is how motivation gains expression in behaviour.

The determinants of motivation and behaviour lie both in the external world and within the animal (see Jensen and Toates, 1997). Stimuli arising in the external world impinge upon the sense organs of an animal and play a role. Behaviour can sometimes be characterized as an automatic response to such stimuli. However, more broadly, it is determined by a range of different types of internal processes, only some of which can be characterized as motivational. The reaction to stimuli can vary as a function of internal physiological events (motivational variables) and such things as learning.

Classification of Behaviour

This section presents a tentative classification of some of the processes that determine behaviour. However, at the outset it is important to emphasize that: (i) the description can only capture a feature of each process, (ii) the distinctions between processes are not absolute, and (iii) there are interactions between these processes. In terms of function, different adaptive problems are reflected in the 'design' considerations underlying the nature of different processes.

Reflexes

Reflexes are somewhat automatic responses to particular stimuli (Gallistel, 1980). They are said to be 'triggered by' the stimulus and consist of single bits of behaviour, often involving only a limited part of the body. Reflexes tend to be rather similar in individuals of a given species. In terms of 'design', they are employed where, in evolutionary terms, a situation has proved to be sufficiently predictable that a single straightforward response can be specified in the context of a given stimulus. Examples are the eye-blink response to an approaching object and the salivation reflex to the presence of palatable food in a dog's mouth.

Modal action patterns

Traditionally, ethologists described so-called fixed action patterns (FAPs) (Tinbergen, 1969). These were seen as stereotyped sequences of behaviour that were triggered by particular stimuli. Further observation revealed that the notion of 'fixed' was a somewhat relative one; there was some flexibility and variation even in FAPs. Hence the term 'fixed-action pattern' gave way to that of 'modal action pattern' (MAP), implying the possibility of some variation around a particular value.

A MAP consists of a *sequence* of behaviour triggered initially by a particular stimulus. Some species-typical mating sequences can be understood in such terms. A 'classical' MAP is illustrated in Fig. 3.1 (Tinbergen, 1969). In this situation, the bird has only a 'rather fixed' sequence of behaviour that it can exhibit. Figure 3.2 shows an attempt at an explanatory model of MAPs. Thus, stimulus 1 (S_1) triggers the initial part of the MAP (response 1; R_1) and behaviour then changes the situation, giving rise to a new stimulus (S_2), which triggers response R_2. This then triggers the next part of the MAP sequence, and so on.

Central pattern generators

Within the central nervous system, there exist systems that generate oscillations of neural activity, termed 'central pattern generators' (Gallistel, 1980). These then influence, amongst other things, motor systems to produce oscillations of motor output. Examples include the oscillations of the legs in walking or the distinctive pattern of scratching shown by the leg of a dog.

Motivation

The term 'motivation' is used here to refer to an internal variable that has a number of effects on behaviour (Toates, 2001). Perhaps the most

Fig. 3.1. A modal action pattern – egg retrieval in the greylag goose.

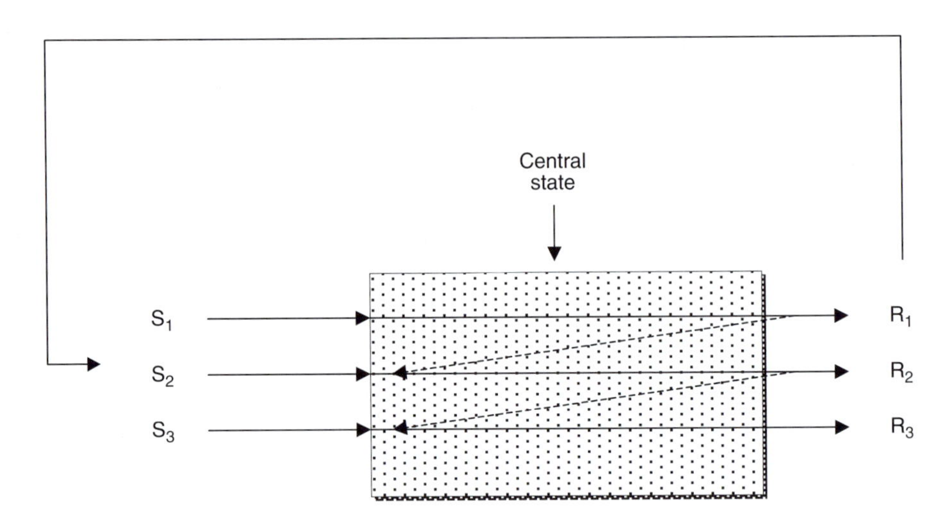

Fig. 3.2. An explanatory model of the modal action pattern. Each stimulus, such as S_1, has a tendency to produce a response, in this case R_1. A central motivational or emotional state (i.e. a particular pattern of neutral activity arising within the brain) modulates the link between stimuli and responses. Performance of a response changes the stimulus array, e.g. R_1 introduces a new stimulus S_2 and tends to excite a response, e.g. R_2. (From Toates, 2001, with permission from Prentice-Hall.)

obvious feature of motivation is that it can vary over time. Motivation is influenced by certain physiological variables of the body. For example, in a state of dehydration, thirst/drinking motivation is normally high. Motivation, as indexed by amount drunk or the work performed to obtain water, increases as a function of deprivation. The hungry animal accepts food, whereas the satiated animal declines it. Thus, in functional terms, motivation is adjusted in accordance with adaptive considerations of 'need'. A motivational system such as hunger/feeding or thirst/drinking is responsible for setting goals. The thirsty animal gives priority to seeking water and engaging in drinking. The motivation that is dominant (e.g. thirst) inhibits the capacity of other motivations (e.g. hunger) to gain expression in behaviour. That is to say, there is a competition between *motivational systems* for the capture of the control of behaviour. There is said to be a final common path (McFarland and Sibly, 1975) and the motivation that gains expression is assumed to do so by capturing this path.

In contrast to the limited range of options of reflexes and MAPs, flexibility can be exhibited by motivational systems and novel solutions to a problem can be found. For example, a rat can employ a wide variety of different techniques for gaining access to food.

As an aspect of motivation, animals are said to form representations of their world, involving 'expectations'. For example, an animal learns the location of food or the expectation that pressing a lever in a so-called Skinner box yields food.

Motivational controls set high-level goals (e.g. get to the food, avoid this predator) and these are then implemented by lower-level processes, right down to individual motor neurons (Gallistel, 1980). This represents 'hierarchical control': the highest level specifies a goal and successively lower levels then implement the command. The lower levels take varying local circumstances into account (Gallistel, 1980). For example, the goal set within the nervous system of a hungry rat might be to get to the end of the maze and obtain food. The goal does not specify exactly which muscles are to be used in achieving this end-point. Tilting the maze slightly means that the same goal can only be achieved by varying the exact motor response employed. Local feedback loops adjust motor output to maintain bodily stability and achieve the goal.

Motivational systems influence other systems just described, such as reflexes, and are influenced by other systems, such as oscillation generators. The next section describes such interdependence between systems.

Producing Adaptive and Coherent Behaviour

Looked at in terms of its immediate causes, behaviour is produced by one whole interdependent nervous system. Hence, any attempt to describe a part of the nervous system as having a particular role in a given behaviour must be interpreted with some caution. However, simply for the convenience of an initial description, it is possible to classify the systems that underlie behaviour, as in the previous section. Having first classified them, in reality they are best studied together, for several reasons. First, to some extent, the systems share responsibility for behaviour and exhibit properties that are complementary. Secondly, there is an interaction between the various types of process. Thirdly, systems are not static but can change their properties over time: behaviour that starts out being characterized as goal-directed can become more reflex-like with repeated experience. This section will illustrate these three features of behavioural control.

Complementary processes

As an example of complementary processes, consider that an animal is confronted by a noxious stimulus (i.e. one that is tending to cause tissue damage) at the skin. There are both reflexive (local, fast reactions of limb withdrawal) and central emotional and motivational controls (involving pain) that serve the function of minimizing damage. As shown in Fig. 3.3, a stimulus that causes tissue damage sets up activity in nociceptive neurons (a nociceptive neuron is one, the tip of which is sensitive to tissue damage such that this sets up action potentials) (Guyton, 1991). This has two effects. By means of a series of interneurons, a link is made with skeletal muscle. A nociceptive reflex produces a motor response that removes the limb from the offending object. This is an automatic, rapid and relatively 'hard-wired' reaction.

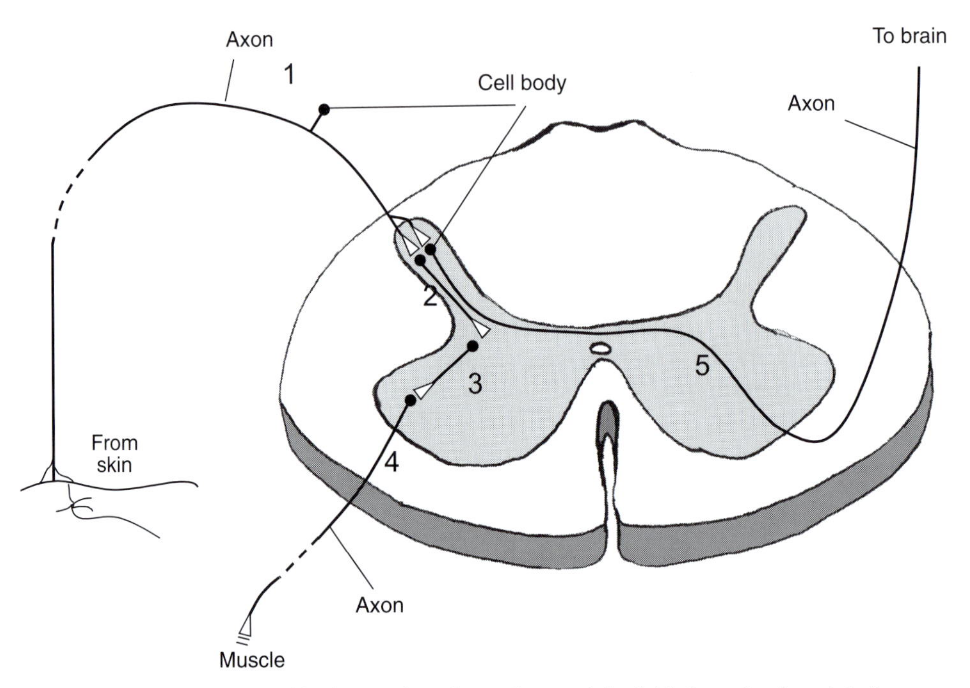

Fig. 3.3. Neurons involved in the nociceptive reflex and the initiation of pain-related behaviour. (From Toates, 2001, with permission from Prentice-Hall.)

Note also the projection of the nociceptive pathway to the brain. It is via input from this pathway that a central emotional state that we term pain is created. Pain has some of the properties of emotional and motivational systems, allowing flexibility of behaviour. Thus, an animal can lick a site of a wound or can place less weight on a painful leg. It might also return to a safe location to allow healing to occur.

Interacting processes

Modulation of reflexes

Figure 3.4 shows an example of the modulation of a reflex by emotion: the startle reaction that an animal exhibits to a loud sound (Davis, 1992). (The term 'modulation' means that the central emotion does not trigger the reflex but alters the strength of any response that a stimulus causes.) This can be described by the term 'hierarchical control': brain regions high in a hierarchy exert control over reflexes organized at a lower level (Gallistel, 1980). The role of hierarchical control makes sense from a functional perspective. This reflex is organized at the level of the brainstem but modulated by a central emotion of fear organized in the brain region that is termed the amygdala: the frightened animal (in comparison with a non-frightened control) shows a stronger response to a given stimulus.

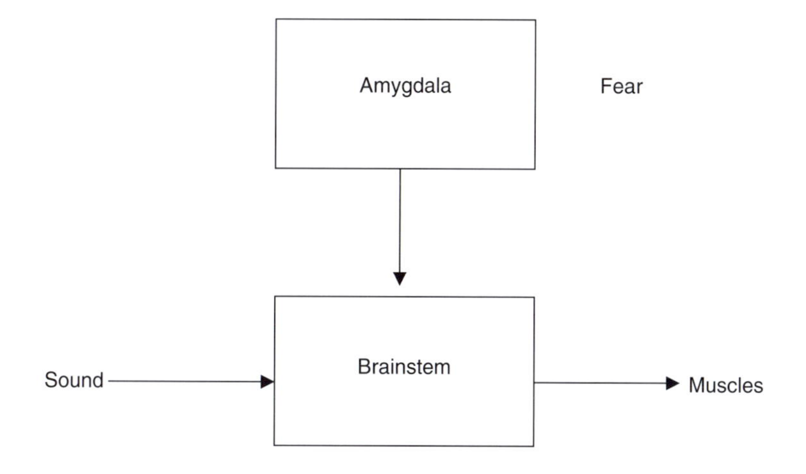

Fig. 3.4. Modulation of a reflex by a central emotional state.

Figure 3.5 represents the system controlling the mating posture (called lordosis) exhibited by sexually receptive female rats (Pfaff, 1989). This posture is triggered by the presence of the male, facilitates intromission and is rather reflexive and stereotyped. It is organized at the level of the spinal cord: there is a spinal circuit of neurons that links the tactile stimulus to a motor response (Pfaff, 1989). However, note the descending pathway from the brain. Via the neurons of this pathway, the reflex is modulated, to make it either more or less likely to occur in response to a tactile stimulus. Activity in this descending pathway depends upon the action of hormones within particular parts of the brain (Pfaff, 1989). When the animal is primed by hormones, such that she is sexually motivated and receptive, the descending pathway modulates the reflex link to make it likely that a tactile stimulus will trigger lordosis. When such hormonal/motivational priming is absent, the same stimulus from the male rat fails to elicit lordosis. Cycles of such sensitization and desensitization correspond to the oestrous cycle. The functional significance is clear: mating is performed at times when the hormonal profile of the female is such that fertilization can occur.

As far as the flexible motivational aspect is concerned, when in a receptive state the female rat shows motivation directed to a male. For example, she can be trained to press a lever in a Skinner box (in response to a press of a lever, a male rat drops down a chute into the Skinner box with the female) or run a maze for such reward. The performance of the appetitive phase of the mating sequence can show considerable flexibility. However, it only occurs when the hormonal state of the female is appropriate for fertilization. It can be seen that there is functional coherence between the motivational and reflex aspects of mating.

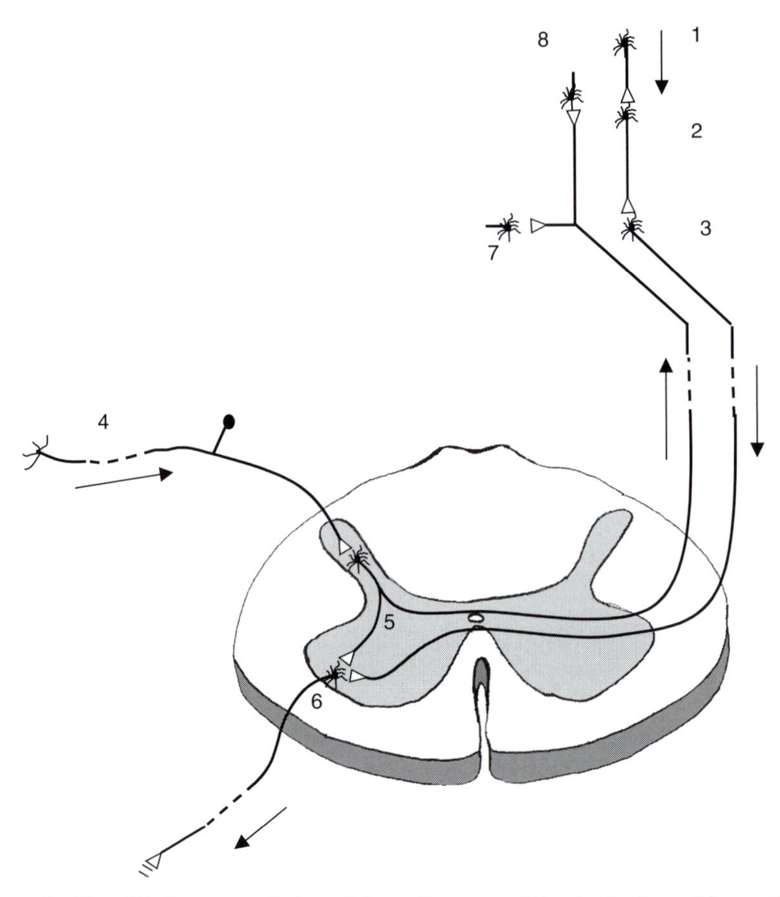

Fig. 3.5. Simplified representation of the reflex organizing lordosis and its modulation by the brain. Neurons 4, 5 and 6 mediate the reflex link between tactile stimulation from the male and the motor output. The pathway 1, 2 and 3 mediates a link from the hypothalamus via the brainstem to modulate the efficacy of the reflex. Ascending information on the tactile stimulus is conveyed via neurons 4 and 5 to influence neurons (7 and 8) at various sites in the brain. (Derived from Pfaff, 1989.)

Motivational control of modal action patterns

As shown in Fig. 3.2, modal action patterns (MAPs) also come under motivational control, as represented by the 'central state'. Thus, in certain motivational states, a MAP is more likely to be shown than in other states. For example, activity within a sexual motivation system would increase the chances that a mating MAP is exhibited.

Rhythms and motivation

Internally generated rhythms interact with motivational states. For example, animals normally exhibit a circadian rhythm of sleep, feeding and drinking, etc. (a circadian rhythm is an internally generated rhythm that goes through a cycle every 24 hours). This points to a rhythmic circadian influence on motivational systems.

Reciprocally, particular motivational systems influence the 'switching on' of particular rhythms. Thus, a thirst motivation and the presence of a water source tend to switch on the lapping pattern of an animal.

Changing system properties over time

The control of behaviour appears to be something of a trade-off between flexibility and stereotyped, automatic responding. Under some circumstances, flexibility and 'open-endedness' are needed, as in exploiting a novel environment. Under more predictable conditions, a rapid automatic response might provide the best solution to the same problem.

Under some circumstances, the parameters of the control of behaviour change (Toates, 2000, 2001). This changes the performance characteristics of the controls and thereby changes the position of behaviour along the flexible–stereotyped axis: behaviour produced initially by a flexible motivational system can start to appear in a 'reflex-like' form or like a modal action pattern. For example, suppose in learning a new task, the environment is sufficiently constant that components of behaviour (B_1, B_2, B_3 ...) regularly follow each other in a sequence. It appears that, as a result of repetition, the performance of B_1 can come to trigger B_2, which in turn triggers B_3 and so on. Transitions are biased by the piece of behaviour just performed. Thus, initially flexible and variable learned behaviour can come to resemble a modal action pattern.

To illustrate a point about changes in the mode of control, let us switch for a moment from non-humans to the human situation. Imagine yourself to be negotiating a novel environment, say, a street covered with ice. Your full conscious control is employed all of the time as you compute your strategy, carefully weighing up the possibilities of motor actions and outcomes of different responses. Contrast that with your regular walking down a familiar street when no ice is around. You are on something like auto-pilot control, with moment-by-moment decisions on walking being made at lower levels of the behavioural hierarchy. Non-humans also appear to devolve control to lower levels when a situation becomes predictable.

Principles and Models of Motivation

Various theoretical models of motivation are available and these represent different traditions. At one time, those representing classical ethology and biopsychology were seen as competitive but more recently some reconciliation has appeared (Toates and Jensen, 1991). First, these two main traditions will be described.

The classical ethological approach

According to classical ethological theory, a particular motivation increases as a function of the length of time that has elapsed since it was last expressed in behaviour (Tinbergen, 1969; Lorenz, 1981). This model seems to offer the prospect of some theoretical integration, pointing to common features between different motivational systems, such as feeding and aggression. According to the model, *specific energy* underlying a motivation accumulates in the absence of behaviour and then discharges with the performance of behaviour.

The model has been criticized on a number of grounds. Thus, there is no evidence of an actual physical accumulation of energy within the animal's body. However, this argument has been answered by the claim that the system acts *as if* there is such an accumulation. Therefore it can serve as a useful metaphor or analogy of how things work. The model has also been criticized on the grounds that it might explain some types of motivation but not all. For example, it would seem to be maladaptive for an aggressive tendency to accumulate as a function of the time since it was last exhibited. From a functional perspective, it would seem to make sense for aggression to be tied to threats arising in the external environment. The criticism has also be made that some types of behaviour, e.g. feeding and drinking, can be explained without the notion of motivational energy. Rather, they can be explained primarily in terms of observable physiology, such as an increasing need for nutrients and increases in loss of water (discussed shortly).

In spite of a number of inadequacies of the classical model, some behaviours do appear to fit its predictions rather well (Vestergaard *et al.*, 1999). The performance of dust-bathing in domestic fowl is one such example. The tendency to perform this behaviour seems to arise as a function of the time spent since it was last performed. In the case of such behaviour, it appears to make sense for there to be an intrinsic timing device with some of the features of motivational energy. In this case, unlike feeding and drinking, there is no obvious homeostatic variable (see next section) that could underlie the basis of motivation.

The biopsychological approach

Within the tradition of biopsychology, the topic of motivation and the associated behaviour tends to be explained by observable physiological events within the animal's body (Toates, 2001). The most successful examples of explanation by this approach concern feeding, drinking and behavioural temperature control.

In the case of drinking and temperature control, there is a physiological variable that is *regulated* by the *control* that is exerted over behaviour. In other words, homeostasis is achieved partly by means of behaviour. Thus, a deviation from optimum in body fluids or body temperature is a powerful determinant of respectively, thirst and control action in the interests of temperature regulation. For example, the hypothermic animal might engage in huddling with other animals or can be persuaded to press a lever to gain units of heat, an index of motivation. The internal body physiology takes corresponding action to correct body temperature, such as shivering. This shows the coherent action between internal physiology and behaviour. Although we might not understand exactly what the determinants of hunger motivation and feeding are, researchers can begin to find some good links with the nutrient needs of the body.

Theories within this tradition emphasize that motivation arises from a combination of external stimuli and internal physiological states. The external stimulus is sometimes termed an 'incentive'. For example, feeding motivation depends upon both the incentive of food and the internal state of nutrient level. The power of the incentive can vary. An animal might decline one food but avidly ingest another. The power of the incentive also depends upon past experience with it. For example, a food that was previously ingested might be rejected if its ingestion was followed by gastrointestinal upset (known as the 'Garcia effect').

Attempts are sometimes made to understand social motivations in terms of homeostatic-like processes. Thus, grooming appears to increase the level of endogenous opioids (morphine-like substances that are produced naturally by the body) and it might be explained by a control system that seeks an optimal level of this physiological variable. Withdrawal from social contact can be followed by a distress syndrome that might be explained partly in terms of biochemical changes ('homeostatic imbalance'). These would motivate the restoration of social contact.

Viewing motivation in these terms gives some unity in comparing motivations that serve homeostasis (e.g. drinking) and those not concerned with physiological regulation (e.g. sexual behaviour and exploratory behaviour). In each case, we can describe motivation and behaviour as being determined by a similar set of factors (Toates, 2001):

1. Physiological states, e.g. hormones in the case of sexual behaviour;
2. External stimuli; and
3. The setting of external stimuli into a context (for example, exploration is particularly triggered by disparity between incoming stimuli and memories of an environment, sexual behaviour can be rearoused in some species by a change of partner).

Viewing motivation in terms of observable physiology highlights the coherence between intrinsic and behavioural controls. For example, insulin is a hormone released in response to the presence of glucose in the blood. It facilitates the transport of glucose across the cell wall and into the cells of the body. Its release under these conditions is somewhat reflex-like. However, the system also shows anticipation. Thus, when an animal such as a rat starts to ingest palatable food there is a release of insulin, termed the 'cephalic phase' of release. This means that glucose can move into the cells in anticipation of the arrival of glucose in the blood and therefore: (i) the cells get their glucose earlier than would otherwise be the case, and (ii) blood glucose level does not rise to a dangerously high level.

An integrative perspective

Traditionally the classical ethological model, rooted in hypothetical internal processes, and the biopsychological model, rooted largely in observable physiology, were seen as competitive for the truth. However, a more recent theoretical development is to argue that both models can coexist, each capturing features of reality (Toates and Jensen, 1991). For example, both point to a combination of external and internal factors as being the source of motivation.

One observation that points to a compromise between different perspectives is that not all feeding and drinking can be accounted for in terms of physiological homeostasis. In addition, there appears to be a motivational tendency that increases simply as a function of time since these behaviours were last performed. Also, the actual performance of feeding behaviour itself seems to play a role in satiety, over and above that of any nutrients ingested. Water obtained by intravenous infusion does not satiate thirst to the same extent as obtaining the same amount of water by mouth. Although there is rather little evidence for a build-up of aggressive tendency as a function of time, none the less, when an animal is placed in a situation of stress, the performance of aggressive behaviour can have consequences for the physiology of the body. For example, in some species, the degree of gastric ulceration can be less if there is an opportunity to exhibit aggression-like behaviour.

Brain Mechanisms and the Control of Behaviour

So far, we have outlined the nature of the controls of behaviour and physiological states. The role of brain mechanisms in such control was implicit throughout. This section takes a closer look at the brain and makes more explicit how it exerts control.

As an organizing framework, it is useful to think of the brain as locked into interaction with the body's external and internal environments. The brain 'looks out' to the external environment and 'looks in' to the internal environment. It computes information on events in both environments and, based on such information, produces adaptive behaviour and adjustments to internal physiology.

To take the example of feeding and nutrient state, we saw in the case of insulin that an external factor, the presence of food, can trigger the secretion of insulin, which affects internal nutrient state. Information on the state of gut contents and nutrient metabolism at the liver is communicated to those brain regions concerned with organizing feeding motivation. This section will look at some examples of how, by taking both internal and external events into account, the brain organizes adaptive behaviour. As such, it will show the interdependence between nervous, endocrine and immune systems.

The brain exerts control over the external environment, mediated via the somatic nervous system and skeletal muscles. It mediates control over the internal environment via, amongst other things, the autonomic nervous system and hormones, smooth muscle and cardiac muscle. There is normally a functional coherence between these two aspects of control. For example, the behaviour of fleeing, mediated by the somatic nervous system, would be accompanied by acceleration of the heart rate, mediated via the autonomic nervous system. This section will look into the nature of such functional coherence.

General principles of control by the brain

The brain is informed of events within the body via the endocrine system, immune system and peripheral parts of the nervous system. Certain hormones released into the bloodstream find their way to the brain where they influence neural events. You saw earlier the example of oestrogen, which affects sexual motivation and receptivity in females. For another example, insulin released by the presence of glucose in the blood appears to play a role in the satiety of feeding and thereby the end of a meal. The hormone cholecystokinin (CCK), released in the gut by the presence of food, also contributes to satiety. This exemplifies negative feedback control. The hormone testosterone released from the testes of the male plays a role in both organizing the motivational systems of mating (in early development) and activating these same mechanisms (in sexual maturity).

For information to be conveyed in the blood and influence the nervous system, then of course it must be capable of reaching neurons. The brain itself is protected from certain events within the blood by the blood–brain barrier. However, there are several ways in which information can pass this barrier. First, some substances within the blood are able to pass the barrier freely. Secondly, in certain parts of the brain there are nuclei at which the blood–brain barrier is relaxed. This allows chemicals in the blood access to particular neurons that are sensitive to particular events. Thirdly, detectors in the peripheral nervous system are sensitive to specific events within the body. For example, products of an immune reaction are detected by the peripheral nervous system and information on the immune system is thus conveyed neurally to the brain.

Figure 3.6 shows a section through the rat brain and highlights some of the regions that are of greatest concern in the present context. Their relevance will be described in the following sections, which look at different examples of control.

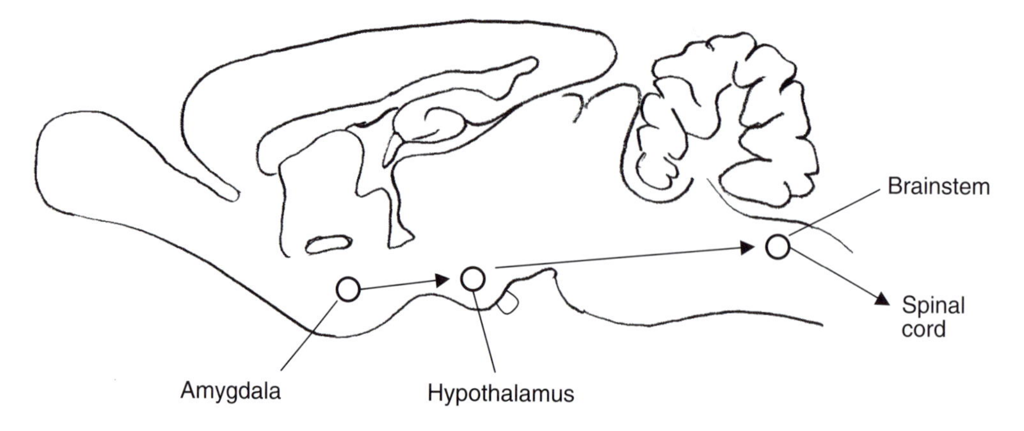

Fig. 3.6. A section through the rat brain, showing location of some structures.

Emotional reactivity and stress

Emotions are processes within the CNS that give a bias towards a range of behavioural options and against another range. Information on the external environment is transferred by various routes to the brain region termed the amygdala (Fig. 3.6), where emotion appears to be attached to it and integration of information occurs. Information is then conveyed to other brain regions, such as the hypothalamus and brainstem. Within this sequence of brain regions, a course of goal-directed behaviour is computed. Finally, by means of descending information, the course of goal-directed behaviour is effected, e.g. attack, flee or freeze.

The autonomic nervous system

In parallel with effecting behaviour, a computation of emotional information is conveyed to the nuclei (e.g. in the brainstem) that organize the activity of the autonomic nervous system (ANS). The computation of an active behavioural strategy of attacking or fleeing would be accompanied by excitation of the sympathetic branch of the ANS and inhibition of the parasympathetic branch. This combination of ANS changes results in the excitation of the heart and the diversion of blood from such places as the gut to the skeletal muscles that are employed in effecting action. There is an increase in the rate of secretion of the hormones adrenaline and noradrenaline from the adrenal gland. There is some controversy on the nature of the behavioural effects of these hormones but information on peripheral noradrenaline is conveyed to the CNS and plays a role in the consolidation of memory.

When the behavioural strategy is one of passivity, there can be a bias towards domination by the parasympathetic branch of the ANS and away from the sympathetic branch. This would be during, say, sleep.

CRF, ACTH and the secretion of corticosteroids

Another biochemical having an important role in the reaction to threat is 'corticotrophin releasing factor' (CRF), sometimes termed corticotrophin-releasing hormone. As shown in Fig. 3.7, this substance serves as a neurohormone (it also serves as a neurotransmitter/neuromodulator within the CNS). As a neurohormone, it is released from neurons with cell bodies in the hypothalamus. It is transported in a local blood vessel the short distance to the pituitary gland, where it triggers the release of adrenocorticotrophic hormone (ACTH) (Fig. 3.7). ACTH circulates in the bloodstream and targets the cortex of the adrenal gland, causing the release of hormones of the class termed 'corticosteroids'. The sequence of hypothalamus → CRF → ACTH → corticosteroids is termed the hypothalamic–pituitary–adrenocortical axis, abbreviated as HPA axis. In its other role, as a neurotransmitter/neuromodulator within the CNS, CRF targets neurons having a role in negative emotion such as fear.

Corticosteroids released from the adrenal gland have a role in energy metabolism. In this, they serve an adaptive function at times of emergency, helping to mobilize energy reserves. As shown in Fig. 3.8, they exert a negative feedback effect on the HPA axis, e.g. at receptors in the hippocampus. When the threat is removed, the HPA axis normally returns to its baseline level of activity. However, protracted and unresolved threat is associated with excessive secretion of corticosteroids, a classical index of stress. Acting via receptors in the brain, e.g. at the hippocampus, corticosteroids probably play a role in modulating behaviour.

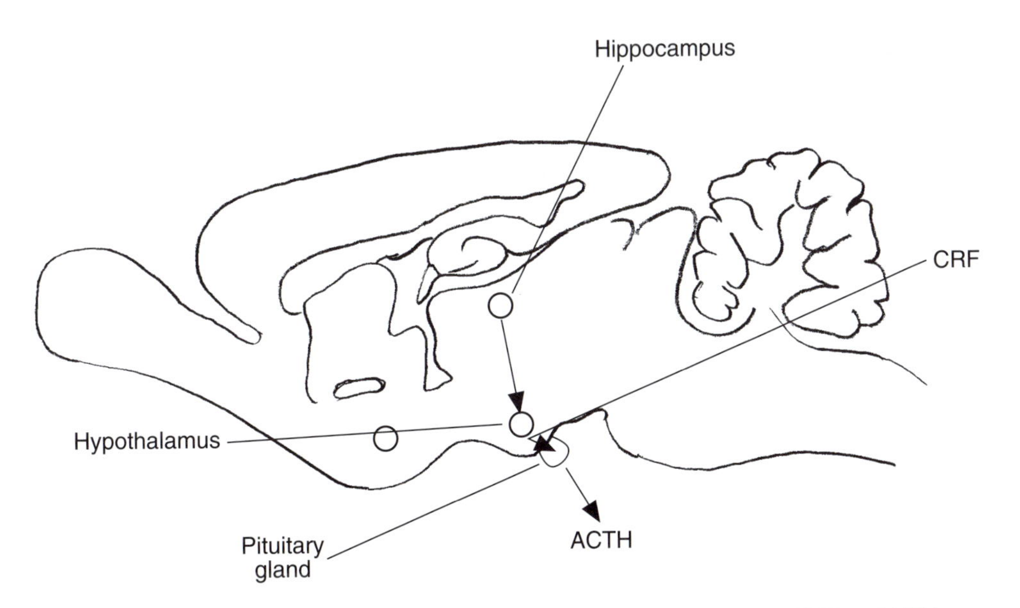

Fig. 3.7. A section through the rat brain, showing the hormonal sequence CRF → ACTH.

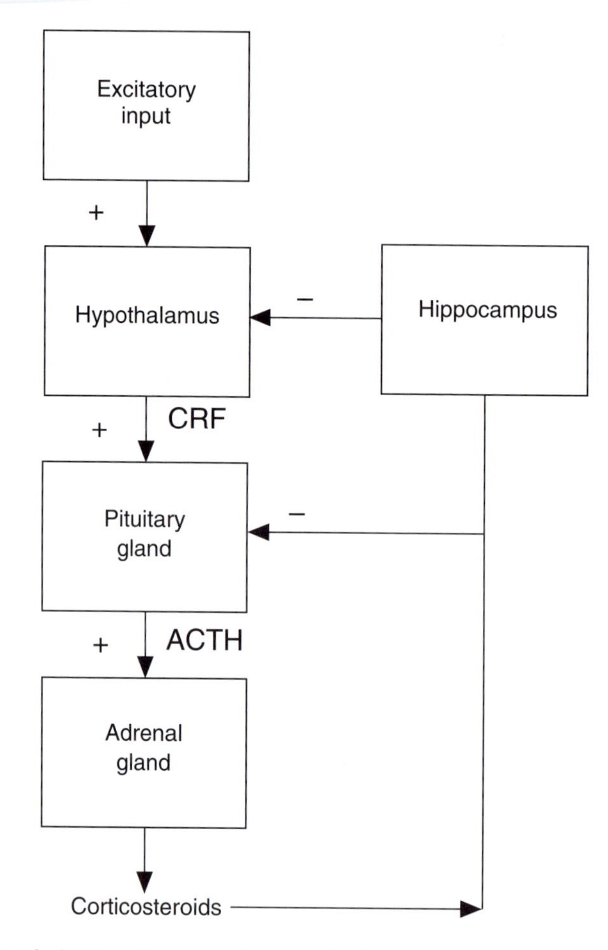

Fig. 3.8. The hypothalamic–pituitary–adrenocortical axis.

Interactions with the immune system

Activity within the nervous system affects the immune system. For example, chronic stress can be associated with a suppression of immune activity. This suppression is mediated via links within the ANS and via the action of corticosteroids. There is some controversy concerning the possible functional significance of such suppression.

Reciprocally, activity within the immune system has effects upon the nervous and endocrine systems. For example, products of an immune response are conveyed via the bloodstream and exert effects upon the CNS. This causes a bias towards inactivity and sleep and the loss of positive motivations of engagement with incentives such as food and a sexual partner. As part of coordinated action, body temperature rises. It is not difficult to appreciate the functional significance of this pattern at a time of, say, bacterial infection.

The brain and homeostasis

The brain monitors directly a number of conditions in the body, such as fluid levels and temperature. Motivation is aroused by a deviation of these from their optimal levels. The brain is also informed of physiological events in the remainder of the body, and these also contribute towards motivation. For example, an increase in central body temperature is detected by neurons in the hypothalamus and action – both behavioural (e.g. seeking a cooler environment) and autonomic (e.g. sweating) – is triggered. However, these actions do not depend simply upon hypothalamic temperature, which is generally well cushioned against environmental temperature stresses. Detection of hypothermia or hyperthermia at the periphery and conveying this information to the hypothalamus will also trigger corrective action.

Some principles of homeostasis can be illustrated by the example of body-fluid regulation, which is considered next.

Body-fluid homeostasis

The defence of body-fluid volume is obviously of crucial importance to survival, and several behavioural and intrinsic physiological processes are organized to defend body-fluid volume. Loss of body fluids is detected by neurons in the brain which convey this information to motivational processes within the hypothalamus and other areas. The detection of dehydration by neurons in the brain triggers the release of a hormone, arginine vasopressin. This acts at the kidney to conserve water.

As another control in the defence of blood volume, loss of fluid from the blood is a potent trigger to the secretion of the substance renin from the kidney. In response to renin, increased production of a hormone, angiotensin ANG II, is stimulated.

ANG II, acting as a hormone, influences the nervous system. That is to say, there are angiotensin-sensitive neurons at locations accessible to blood hormones. The blood–brain barrier is relaxed at certain sites (circumventricular organs) where angiotensin-sensitive neurons are located. These neurons provide information to brain regions that organize drinking motivation, such as the hypothalamus. A powerful stimulus to drinking is the injection of ANG II into certain regions of the brain. This causes an animal to stop what it is doing and switch to drinking. The assumption is that the substance triggers thirst motivation.

Secreted in response to the loss of blood, ANG II has effects on both: (i) intrinsic physiological regulation (e.g. blood pressure, retention of water and sodium by the kidney), and (ii) behaviour, in the form of increasing motivation directed to water and sodium incentives. This represents coordinated action in that both the behavioural and intrinsic physiological effects serve body-fluid homeostasis.

The role of a sex hormone: testosterone

Testosterone is released from the testes of the male in response to a hormonal sequence ('axis') comparable to that just described for the HPA axis (testosterone is also present to some extent in females, being released from the adrenal gland). Depending upon the species, it can either be released at a relatively constant rate or there can be cycles of release corresponding to the pattern of reproduction. Acting on the CNS, testosterone sensitizes both sexual motivation, and, in some species, the tendency to aggression.

Testosterone is a good example of the dynamic nature of the interaction between internal and external factors controlling behaviour. Testosterone influences behaviour via the CNS, but, reciprocally, the consequences of behaviour influence testosterone. Thus, a history of winning fights tends to increase the secretion of testosterone, thereby consolidating the aggressive strategy. Conversely, defeat or loss of status within a group tends to cause a reduction in testosterone secretion, which biases towards a passive strategy of, say, submission.

The Relevance of a Motivational Perspective to Applied Ethology

So much for the more 'pure' aspects of motivation, the brain and the controls of behaviour, what does all this imply for applied ethology? The topic of motivation is often to the fore in discussions of animal welfare. This section gives some examples of where motivational ideas, as discussed earlier, are of relevance to issues of welfare.

Satiety of behaviour

Applied to welfare issues, the classical ethological model appears to suggest that it is important for animals to be allowed to exhibit species-typical behaviour. Some qualification is suggested here. For example, expressed in terms of the model, there is no evidence that a tendency to perform escape behaviour increases as a function of time since it was last shown. This implies that it would, of course, be illogical to argue that the welfare of a prey species is really improved by providing the opportunity for escape behaviour and therefore presenting predators at intervals. However, there can be more subtle examples where the predictions of the classical ethological model might need to be taken into account. Some feeding systems do not permit domestic animals to perform much natural feeding behaviour in the gain of nutrients. This could mean that normal satiety is not attained and the digestive sequence (e.g. release of hormones) is not properly triggered. In nesting situations, providing the opportunity for the animal to perform nesting behaviour might improve welfare relative to simply providing a completed nest.

Stereotypies

Stereotypies are sometimes described as repetitive and apparently pointless behaviour (see Chapter 6 for further discussion). It is unclear what is the motivational basis for their appearance, or if even motivation is a useful concept in understanding them. Another point of discussion is whether there are consequences of performing stereotypies, for example, such that welfare is improved.

The perspective developed here suggests another interpretation: stereotypies might be the inevitable outcome of a restricted environment (Toates, 1995, 2000). If the barren environment provides insufficient opportunities to perform flexible and variable sequences of goal-directed behaviour, then more stereotyped controls lower in the hierarchy might automatically gain control. Thus, physically present stimuli (with which the animal is in close and permanent contact) could get locked into association with particular responses, which become self-reinforcing. Transition biasing could mean that performing one part of a sequence (behaviour 1) triggers another (behaviour 2) and behaviour 2 then automatically triggers behaviour 1, and so on.

Goals and expectations

By attributing such things as goals and cognitions to animals, the behavioural scientist is led to a richer concept of the needs of a captive animal. Animals are seen to have not only needs for such things as food, water and shelter and avoidance of tissue damage and pain, but also have cognitive needs (or 'wants'), such as the avoidance of excessive frustration or boredom.

References

Davis, M. (1992) Analysis of aversive memories using the fear-potentiated startle paradigm. In: Squire, L.R. and Butters, N. (eds) *Neuropsychology of Memory.* The Guilford Press, New York, pp. 470–484.

Gallistel, C.R. (1980) *The Organization of Action – a New Synthesis.* Lawrence Erlbaum, Hillsdale, New Jersey.

Guyton, A.C. (1991) *Textbook of Medical Physiology.* W.B. Saunders, Philadelphia.

Jensen, P. and Toates, F.M. (1997) Stress as a state of motivational systems. *Applied Animal Behaviour Science* 53, 145–156.

Lorenz, K.Z. (1981) *The Foundations of Ethology.* Springer-Verlag, New York.

McFarland, D.J. and Sibly, R.M. (1975) The behavioural final common path. *Philosophical Transactions of the Royal Society of London B* 270, 265–293.

Pfaff, D.W. (1989) Features of a hormone-driven defined neural circuit for a mammalian behaviour. *Annals of the New York Academy of Sciences* 563, 131–147.

Tinbergen, N. (1969) *The Study of Instinct.* Clarendon Press, Oxford.
Toates, F. (1995) *Stress – Conceptual and Biological Aspects.* John Wiley & Sons, Chichester, UK.
Toates, F. (2000) Multiple factors controlling behaviour: implications for stress and welfare. In: Moberg, G.P. and Mench, J.A. (eds) *The Biology of Animal Stress.* CAB International, Wallingford, UK, pp. 199–226.
Toates, F. (2001) *Biological Psychology: an Integrative Approach.* Prentice-Hall, Harlow, UK.
Toates, F. and Jensen, P. (1991) Ethological and psychological models of motivation – towards a synthesis. In: Meyer, J.A. and Wilson, S. (eds) *From Animals to Animats.* MIT Press, Cambridge, Massachusetts, pp. 194–205.
Vestergaard, K.S., Damm, B.I., Abbott, U.K. and Bildsøe, M. (1999) Regulation of dustbathing in feathered and featherless domestic chicks: the Lorenzian model revisited. *Animal Behaviour* 58, 1017–1025.

Learning and Cognition

4

Björn A. Forkman

Any animal that can predict the future has a tremendous advantage over one that cannot. Predicting, and to some extent controlling the future, is really what learning is all about. Through learning the animal gets the chance to respond to cues, e.g. the hen to react to the scratching sound of the beetle, before the relevant event happens – the beetle emerges and can be caught. It can even learn to modify its behaviour so that pleasant things happen and unpleasant things do not. Understanding the function of learning, that it helps animals to predict and control future events, will make it much easier to understand and remember the different principles found in this chapter.

Cognition

Cognition has been defined as psychic processes that cannot be observed directly but for which there is scientific evidence from which they can be inferred. However, the problem is that scientists vary widely in what they consider to be scientific evidence, and therefore in which cognitive capabilities they assign to animals. The stance taken in this chapter is a sceptical one: it is a sound scientific principle to always prefer the simplest explanation when more than one is offered. In psychology a special case of this principle was formulated by Lloyd Morgan and is therefore known as Lloyd Morgan's canon: 'In no case may we interpret an action as the outcome of the exercise of a higher psychical faculty, if it can be interpreted as the outcome of one which stands lower in the psychological scale' (in Roberts, 2000).

The question of awareness or consciousness is one that is often brought up when speaking about cognition. The most commonsense explanation is that animals, at least vertebrates, are conscious of what is happening around them. This does not constitute a valid scientific reason for assuming that that is the case, however. In practice it has been shown to be well-nigh impossible to define and/or demonstrate consciousness in animals, and today scientists range from those stating that a wide range of animals are conscious (Griffin, 1992) to those stating that only humans are (MacPhail, 1998). Among cognitive scientists, perhaps with the exception of primatologists, there is now a tendency to concentrate on how information is encoded and used rather than on the question of awareness.

It is all too easy to equate complex behaviours with complex cognitive or mental abilities. One famous example showing the dangers of this is the case of the hygienic and unhygienic bee strains. The hygienic bee strain uncaps cells with dead larvae, and removes these larvae from the hive, while the unhygienic strain does not. The hygienic behaviour results in the removal of dead larvae from the hive, and it might be tempting to say that this is the goal of the individual bee. Crossing and then back-crossing the two strains results in bees with different behaviour patterns. Some bees behave like the original hygienic and unhygienic strain, some of the offspring only uncap, but do not remove the larvae, and some do not uncap the cells, but will remove any dead larvae if the cells are uncapped for them (Rothenbuhler, 1964). The fact that 'half' of the hygienic behaviour can exist (and that it does not fulfil any purpose on its own) clearly argues against any explanation involving very complex goal representations.

It is important to remember that the fact that there is no scientific evidence for a given capability does not mean that the animal does not have it. We have, for example, no scientific evidence for self-awareness in chickens – this is not to say that they do not have it, just that we have not been able to show it. This distinction between what is scientifically proven now and what may be proven in the future is especially important to make when dealing with questions of animal welfare. To give the animals 'the benefit of the doubt' is not a valid scientific argument but can be a humanitarian one.

One-event Learning

One-event learning is the learning process in which the reaction of an animal to an event is dependent on the fact that it has encountered an event previously. (There is no association between two specific events, as there would be for associative learning.) Types of one-event learning include habituation and sensitization.

Habituation is the simplest type of learning, and one that seems ubiquitous in the animal kingdom. The definition of habituation is a decrease in an innate response caused by repeated presentations of the same stimulus. Habituation can occur in two slightly different ways. For the first type, take as an example the first time a horse hears the wind rustling in the leaves. This may cause it to flee, but if the rustling happens often enough, and nothing else occurs, the horse will stop responding to the rustling. Functionally this means that if a given stimulus is not followed by an important event (in the case above, e.g. wolves bursting out of the woods), the animal stops attending to that stimulus. This type of learning shares many similarities with the phenomenon of extinction discussed below.

A slightly different type of habituation is the decrease in responding that occurs when a hedonic stimulus (hedonic means that the stimulus causes an emotional aspect on a scale from pleasure to displeasure) or a resource is continuously present, e.g. food for laying hens fed *ad libitum*. The easiest way of demonstrating the presence of this is by breaking the habituation, i.e. dishabituate the animal. If we take a hen that has free access to food, and measure the amount of feeding it performs during 10 minutes, and then measure the feeding for 1 minute after having stirred the food, we will normally find an increase in the feeding behaviour after stirring. It is as if we had brought the food to the attention of the hen again. (The habituation of the horse in the previous example can of course also be broken, e.g. by an attack. The horse then starts responding to the rustling as if no habituation had occurred.)

Sensitization is another type of one-event learning. In sensitization the animal encounters a hedonically laden stimulus or situation, e.g. is frightened by a predator; this makes the animal more likely to react to any new stimulus as if it was a predictor of a new attack. The same type of increased sensitivity can occur for many other motivational systems, e.g. foraging. A rooster that has recently encountered a beetle will tend to react to any new sound – or sudden movement – as if it was a sign that there is more food around. In the same way, a rooster that has just been attacked will react to the same sounds as if it was a predator.

In a novel or slightly stressful environment the animal will often initially react with sensitization, with the animal reacting very strongly to any change in the environment. If the level of stress is low enough, the animal may subsequently gradually come to habituate to the new environment.

Associative Learning

In associative learning the animal associates one event with another event. In classical conditioning the association is between two events, whereas in instrumental conditioning the association occurs between a behaviour and its consequences.

Classical conditioning

A good blackbird is a blackbird that knows where to look for worms. To be able to predict where there are worms and where there are not, the blackbird has to be able to associate two events with each other, in this case the features of the habitat and the presence or absence of worms. This type of learning, in which the animal associates two events, is called classical or Pavlovian conditioning, and was first studied by the Russian physiologist Ivan Pavlov with his famous salivating dogs (Fig. 4.1).

Fig. 4.1. The automatic or innate response to food is salivation; salivation is therefore an unconditioned response (UR), and the food is an unconditioned stimulus (US). By ringing a bell just before the presentation of the food, the bell and the food become associated. After a number of trials the dog will start salivating as soon as it hears the bell. The bell has become a conditioned stimulus (CS), while the salivation is now a conditioned response (CR). The same behaviour, salivation, can be either a UR, when it is a reaction to a US, or a CR, when it is a reaction to a CS.

The animal associates one event (which will become the conditioned stimulus, the CS) with another event (the unconditioned stimulus, the US). The association is detected when the animal starts to react to the conditioned stimulus as if it was the unconditioned stimulus (but see below). Pavlov's dog starts to salivate when it hears the bell (a conditioned response, CR, since it is a reaction to a conditioned stimulus), just as it salivated at the sight of the food (unconditioned response, UR, since it is an innate reaction to food). It is important to remember that what the dog has learned is an association between the bell and the food (Pavlov, 1927, in Domjan, 1998).

There are two possible alternatives for how the association might occur. The first is that the conditioned stimulus evokes the representation of the unconditioned stimulus and that this, in turn, elicits the response; the bell evokes the idea of the food and it is the idea of the food that causes the salivation (CS–US–R, a stimulus–stimulus association). The second alternative is that the conditioned stimulus itself begins to elicit the same response as the unconditioned stimulus; the bell has taken on the properties of food and therefore causes the salivation by itself (CS–R, a stimulus–response association). In the first case the animal in some sense 'knows' that, for example, the bell predicts food (although it is impossible to say exactly how this knowledge is represented or if the animal is aware of it). In the second case, however, the animal just reacts to the bell in exactly the same way as it has reacted to the presentation of the food itself in the past, and it is not necessary to postulate any knowledge in the dog to explain its behaviour. The latter is a more behaviouristic explanation, whereas the former is a more cognitive one (behaviourism is a school of experimental psychology that, in its original form, focused exclusively on what could be observed, i.e. the behaviour of the animal; see also Chapter 1).

A number of experiments show that a CS–US–R association can be formed. Holland and Straub (1979) conditioned rats to associate a burst of noise with a new sort of food dropping down into a food trough – the unconditioned response to food is to approach it, so the animals started to approach the food trough as soon as they heard the tone. The animals were then given the new food in their home cage and thereafter injected with lithium chloride (LiCl), a drug that induces nausea. (This treatment ensures that the rats will refuse to eat the food when they encounter it again, see the paragraph on predispositions below.) The rats were then taken back to the tone chamber and exposed to a burst of noise. The reaction of the rats is thought to depend on the type of association they have acquired. If they have formed a CS–R association, the devaluation of the food should not affect the behaviour of the animal. The noise should evoke the behaviour directly. If, on the other hand, the animal has formed a CS–US–R, then the tone should evoke the idea of the food, and since that specific type of food now is not considered edible, it should not evoke the approach response. This latter alternative is in fact what happened. There are very few studies on farm animals, but it seems likely that the mammalian species can form this type of association; there is also recent evidence that this might be true for the domestic hen (Forkman, 2001; see also Regolin *et al.*, 1995).

This devaluation study, and others like it, shows one of the main advantages of a CS–US–R association. The animal is able to vary its behaviour in a very flexible way and even perform an adaptive response in a situation it has never encountered before.

That animals are able to imagine just what CS predicts also has other consequences. In a very early experiment, monkeys were required to select one of two food wells – in most cases they were rewarded with a banana but occasionally they were given a lettuce leaf instead (monkeys like bananas much more than they like lettuce). In most cases the monkeys refused to touch the lettuce leaf, spent a lot of time looking in other places and sometimes flew into a rage and started shrieking at the observers. The ability to have an expectancy makes it possible for an animal to become disappointed or frustrated when that expectancy is not fulfilled. This frustration is not exclusive to higher primates, however; for example, depriving chickens of expected food will also lead to increased activity and aggression (Haskell *et al.*, 2000).

The CS–R hypothesis is more parsimonious than the CS–US–R hypothesis, but it is also a less intuitive explanation – it is hard to imagine that Pavlov's dog does not know that the bell predicts the food. However, there are studies showing that CS–R associations are sometimes formed. Indeed, it seems probable that, at least among birds and mammals, both a CS–R and a CS–US–R association is formed and that it is the relative strength of these two associations that determines the behaviour of the animal. Which association is stronger will depend on a number of things, including the species and the specific circumstances, and therefore has to be determined for each specific case. Generally speaking, however, it seems that prolonged training strengthens the CS–R association.

Classical conditioning is very widespread, both between species and within species. It probably helps the animal to decide, for example, what stimuli are considered alarming, or what to eat (Provenza and Balph, 1987). It may also greatly affect the other type of associative learning – instrumental conditioning.

Instrumental conditioning

A squirrel that wants to open a hazelnut will gnaw at it. If it has never encountered a hazelnut before, the gnawing will be random and it will take a long time for the squirrel to get at the kernel of the nut. However, an experienced squirrel will crack the nut very rapidly and with a minimum of effort. Clearly, the squirrel has learned something about the consequences of its behaviour, and is changing its behaviour based on those consequences.

Whereas in classical conditioning the animal associates two events, two stimuli, with each other, in instrumental conditioning the behaviour of an animal is dependent on the previous outcomes of that behaviour (this type of learning is also known as operant conditioning or trial-and-error learning). If the outcome is desirable, the animal will perform more of the behaviour. If, on the other hand, the outcome of the behaviour is aversive, the animal will perform less of that behaviour (Domjan, 1998) (Fig. 4.2). There are four different possibilities (Table 4.1).

The behaviour itself is often insufficient to result in the following event, it is rather the performance of the behaviour in the presence of a discriminative stimulus that results in the event. The behaviour of sitting does not in itself result in getting a biscuit; however, sitting when someone says ‘sit’ does – if you are a dog that is. The discriminative stimulus can be a single stimulus (like ‘sit’) but it can also be the whole context in which the conditioning takes place, e.g. the place, the smell and the sounds. This means that if the conditioning always occurs in the same place, e.g. you are always training your dog to sit while indoors, your dog will be less likely to obey you outdoors. Not because it is disobedient, but simply because part of the discriminative stimulus that tells it that performing the behaviour will result in a reward is missing.

As we have seen earlier, Pavlovian conditioning can occur either through CS–R or CS–US–R. A similar distinction can be made between

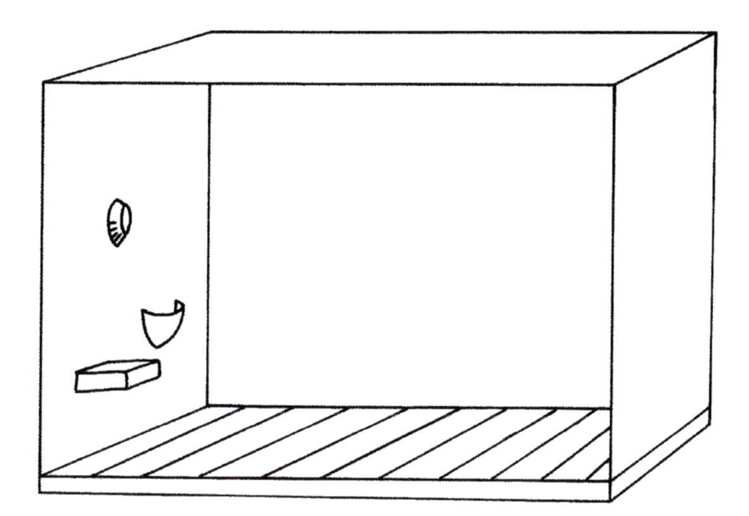

Fig. 4.2. Instrumental conditioning is often investigated in what is known as a Skinner box or operant conditioning box. In a typical Skinner box there is a buzzer to give an auditory signal, a lamp to give a visual signal and a lever to allow the rat (which is the species most commonly used) to make a response. There is also either a food trough or a water nipple, so that positive reinforcements can be given. Finally the animal is standing on a grid that can be used to give a weak shock to the paws of the animal.

Table 4.1. Instrumental conditioning procedures.

Name	Consequence	Result	Example
Positive reinforcement	Response produces desirable consequence	Increase in behaviour	When the dog sits, it gets a biscuit
Positive punishment	Response produces aversive consequence	Decrease in behaviour	When the dog bites, it gets yelled at
Negative reinforcement	Response stops or prevents aversive consequence	Increase in behaviour	If the dog shows submissive behaviour, it doesn't get bitten
Omission training	Response stops or prevents desirable consequence	Decrease in behaviour	If the dog barks, it doesn't get a biscuit

two types of instrumental conditioning: the S–R (seeing a lever causes the rat to press on it) or S–R–US (the rat knows that pressing the lever will result in food). In the S–R–US case the animal is actually able to predict the outcome of a behaviour it has not yet performed. This means that if, and only if, a species is capable of a S–R–US it can be said to have a goal with its behaviour. There is no way of knowing whether this prediction or goal by the animal takes the form of thoughts identical to our own, but the similarities are obvious.

In operant conditioning some Pavlovian conditioning is happening – especially when the animal is attending to the discriminatory stimulus. In some situations this may result in the animal reacting in a Pavlovian instead of an instrumental way. An example of this is when you are trying to teach a dog to run away from you when you say 'forward' by giving it a snack when it performs the behaviour. The dog will learn both the association between its behaviour and the reinforcer, but also form an association between the word 'forward' and the reinforcer (and of course yourself and the reinforcer). This means that there is one tendency for the dog to run away from you since it is only then that he gets a reward, but also a second opposite tendency to not run away. This second tendency is there because the innate response to food is to approach it, not run away from it, and after conditioning it becomes the CR to the CS, i.e. the word 'forward', and yourself. What the dog does will depend on the relative strength of the two processes. In most cases, the behaviour of the animal is much closer in time to the reward than is the discriminative stimulus (e.g. the running away will always be closer to the reward than is the word 'forward' since the running only occurs as a response to the command). If the dog is very hungry (i.e. the reinforcing property of the reward is very strong), you might see the dog hesitating to run away from you, or running only a short distance, even though you would have thought that it would be even more eager to perform the task and get the food.

Many of the problems you might encounter while trying to teach animals different tasks occur because of the conflict between the results of Pavlovian and operant conditioning. To enhance the effectiveness of any task that requires learning (e.g. automatic milking, transponder feeding, etc.) it is worthwhile considering both the classical conditioning and the instrumental conditioning aspects of the task. What you should be asking yourself is whether the behavioural change you are trying to condition is compatible with the UR to the reinforcer you are offering.

A type of instrumental conditioning that deserves special attention is the case of active avoidance behaviours. In active avoidance situations, performing a given behaviour either stops or prevents an aversive stimulus from occurring (i.e. the behaviour is negatively reinforced). Unfortunately, many species will rapidly learn to use aggressive behaviours as avoidance responses. A rat (or sow, horse or even rooster) will very rapidly learn to threaten and even attack you if you inadvertently teach it that attack is an appropriate avoidance behaviour. You do this by responding to the attack by hesitating. Not only is active avoidance conditioning rapidly established, unfortunately it is also typically very resistant to extinction!

Extinction

Extinction is a special type of learning that occurs when a stimulus or behaviour is no longer followed by a reinforcer. The first reaction to this is often a burst of activity but after a while the animal gradually stops responding to the stimulus. The animal does not forget what it has learned; instead it learns that the previous association no longer holds. The fact that the animal does not forget can be clearly seen in the phenomenon of spontaneous recovery. If a hen that has learned to associate a scratching sound with a beetle one day is exposed to a number of scratching sounds without ever finding a beetle, it will stop rushing towards the sound. On first hearing the scratching on a subsequent day, however, the hen will once more rush towards the sound. The extinction will be more rapid on the second day, and even more rapid on the third, and so on.

Not all associations are equally easy to extinguish. The ease of extinction depends on many things and is, to some extent, species specific (see the section on predispositions, below). In general, however, animals that have been trained on a schedule that gives reinforcement in a random manner are more resistant to extinction than those trained on a continuous reinforcement schedule (this is known as the partial reinforcement extinction effect or PREE). To train a dog not to beg at the table, when it has received a reward only now and then, is very difficult! Unfortunately there is no sense in starting to give it a reward the whole time, planning to extinguish the behaviour at a later stage – once the animal has experienced the partial reinforcement schedule the resistance to extinction is there. It is as if the animal does not just learn about the association between the two events, but also about the variability in that association. If the association is very variable, it will be difficult to detect any change in the association, and hence that it has changed. The PREE is not just a problem however, it can also be put to good use if you want an animal to continue performing a response in the absence of reinforcers, e.g. in a competition.

Another type of association that is difficult to extinguish is the active avoidance behaviour discussed above.

Factors Affecting Learning

As stated in the introduction, the function of learning is to make it possible for the animal to predict or manipulate the future. Imagine a hen that has just seen and eaten a fat worm. Every animal experiences an almost infinite number of stimuli at any given time. How can the hen learn how to get more worms? (For a summary see Fig. 4.3.)

Timing

The first factor is the timing of the events. Animals and humans form an association between the reinforcer and an event that occurred a very

Fig. 4.3. A learning story: any hen will approach (*UR*) a worm-like object (*US*). However, the hen in the picture will also approach (*CR*) any dark leaves (*CS*) because it has learned (*Pavlovian conditioning*) that there are worms (*US*) under them. The hen has also learned to attack only the front part of the worm (*instrumental conditioning*). The front part of the worm tastes good (*positive reinforcement*) but it has a sting in its tail (*negative reinforcement*). Because there are worms under the leaves only sometimes (VI/VR), the hen will continue to search for them for a very long time (*PREE*), even when all the worms have disappeared. Eventually, however, the hen will stop approaching the dark leaves (*extinction*).

short time before. If a hen hears a scratching sound just before it finds the worm, an association is rapidly formed, if on the other hand the sound of the worm comes 60 seconds before the hen finds it, the association will be formed much more slowly, if at all. If the sound occurs at the same time as the hen finds the worm, the association is also slow to be formed. This makes sense from a functional point of view, since the function of learning is to predict future events, and events that occur together are, by definition, simultaneous.

Strength and salience of reinforcer

The second factor is the strength or the salience of the reinforcer. An association is more easily formed if the strength of the reinforcer is large (i.e. the worm is fat and juicy) than if it is small. This makes sense since, from an evolutionary point of view, it is more relevant to be able to predict a very important event than a less important one.

Strength and salience of stimulus

The third factor is the strength or salience of the predictive stimulus. The stronger the stimulus, the stronger the association will be. Some animal species pay more attention to certain modalities than others. Which stimulus is the most salient is therefore not dependent solely on the physical strength of the stimulus but also on the particular species and what the US is (see below). The salience of the stimulus is also affected by the frequency with which the animal has experienced it previously, i.e. of how surprising it is (just as with the reinforcer). If the stimulus is completely novel, e.g. a sound that has never been heard before, then the animal will attend more to it and therefore form a stronger association between it and subsequent events. If, on the other hand, you are always whistling when you are walking with your dog, you will have a hard time teaching the dog to attend to a specific whistle signal and learn what that means. The animal has already learnt that the whistling does not hold a predictive value; this is an example of learned irrelevance. In more general terms, learned irrelevance occurs when an animal learns that a given stimulus does not have any predictive power.

Predispositions

The fourth factor concerns the innate learning preferences or predispositions of the animal. Not all stimuli can be equally well associated with a given reinforcer. One of the most striking examples of predispositions in conditioning is a series of experiments done by John Garcia and co-workers in the 1960s and 1970s. In his classic study, rats were given water to drink. The water was either flavoured or had a clicker connected to it so that whenever the rat drank it heard a clicking sound. Drinking the water could result in either the rat receiving a shock or experiencing nausea. The rat easily learned to associate sound and pain, but not sound and nausea. Rats also learned to associate taste and nausea but not taste and pain (Garcia and Koelling, 1966). It is as if animals (and humans) had a given connection already formed between nausea and a smell or a taste – all we need to learn is which smell or taste. In the same way, we have a connection formed between pain and a sound – we just have to learn which sound. The salience of a given stimulus is therefore not only dependent on the timing and 'surprise factor' mentioned above, but also on the type of reinforcer it is to be associated with.

In an anti-predator situation it is very hard to instrumentally condition an animal to perform a behaviour that is not in its 'anti-predator behaviour repertoire'. This makes functional sense – trial-and-error learning with successive approximations is not the best way of dealing with an attacking tiger. The reason for this is that the signal predicting the attack becomes associated with the attack and therefore elicits a conditioned response via Pavlovian rather than instrumental conditioning. This is not to say that it is impossible to change the behaviour instrumentally, just that it is very difficult.

Using associative learning

Associative learning is a very powerful tool to use to change the behaviour of animals. The problem is, as so often, to translate the theory into practice. A good book that can be of some help is one by Pamela Reid (1996). Although the book deals primarily with dog training, the same principles apply to all our domestic animals.

Social Learning and Related Phenomena

True imitation, i.e. that animals are learning completely new behaviours through imitating those of other animals, is probably quite rare. In most cases the behaviour can be explained instead by a tendency of animals to behave towards the same parts of the environment as other animals (local or stimulus enhancement), or by just repeating the same well-known behaviour (social facilitation) (Table 4.2). If one pig starts feeding, others will soon follow suit (social facilitation). If one hen starts feeding from a bowl with flowers other hens will prefer to feed from the same type of bowl (stimulus enhancement). These mechanisms ensure that the behaviours in a social group are synchronized, but can also function to transmit, for example, food preferences. While it is a good idea to be sceptical towards reports of the social transmission of new behaviours, there are some types of social transmission of information that are well documented and important (Shettleworth, 2000).

Table 4.2. Three types of social learning and related phenomena. This is a classification based on the behaviour of the observer animal in relation to the situation.

Name	New behaviour	Repeats behaviour	Acts towards the same type of stimulus	Example
True imitation	X	X	X	Orang-utans will attempt to make a fire after watching humans making fires
Social facilitation		X		You yawn, and everyone else starts yawning
Local enhancement			X	Ducks will land where there are ducks already present

Alarm calling and mobbing, learning what is dangerous

In many species, individuals learn to recognize predators through the reaction of conspecifics. The phenomenon has been seen and studied in monkeys, a number of different bird species, but also in fish. The observer animals often have a strong predisposition (see above) for what they should be fearful of: monkeys will rapidly learn to avoid a snake through observing the alarm reactions of other individuals. When a flower was used instead of the snake no such learning took place. This type of learning seems to follow the same pattern as that of more classical associative learning discussed later.

Food preference transmission, learning what is safe

One of the many problems an animal faces is what to eat. The transmission of food preferences, especially in rats, has been studied extensively. One way in which food preferences can be learned is through local enhancement, as discussed before. However, rats also learn what food to eat in another way. When a rat feeds, small food particles will get stuck on its whiskers and around its mouth. The smell of these food particles, together with the breath of the rat, is very attractive for other rats. They approach and sniff any returning individual around its mouth, and subsequently show a strong preference for that food (Galef, 1996). While it is well known that other mammals, e.g. sheep (Provenza and Balph, 1987; Thorvaldsdottir *et al.*, 1990), can learn what to eat through social learning, much less is known about the exact mechanism through which this occurs.

Learning in an Applied Context

Knowledge of animal learning and the mechanisms involved are of great importance to a variety of aspects on how domestic animals are kept and handled. It allows us to understand what kind of technical equipment animals can learn to operate, for example feed dispensers and water devices. It may help us to find out how animals learn and adapt to new environments and husbandry routines. Furthermore, it provides insights into the cognitive capabilities of an animal, which are essential for assessing the welfare of animals in different situations.

References

Domjan, M. (1998) *The Principles of Learning and Behaviour*, 4th edn. Brooks/Cole, London.

Forkman, B. (2001) Domestic hens have declarative representations. *Animal Cognition* 3, 135–137.

Galef, B.G. Jr (1996) Social enhancement of food preferences in Norway rats: a brief review. In: Heyes, C.M. and Galef, B.G. Jr (eds) *Social Learning in Animals.* Academic Press, London.

Garcia, J. and Koelling, R.A. (1966) Relation of cue to consequences in avoidance learning. *Psychonomic Science* 4, 123–124.

Griffin, D.M. (1992) *Animal Minds.* University of Chicago Press, Chicago.

Haskell, M., Coerse, N.C.A. and Forkman, B. (2000) Frustration-induced aggression in the domestic hen: the effect of thwarting access to food and water on aggressive responses and subsequent approach tendencies. *Behaviour* 137, 531–546.

Holland, P.C. and Straub, J.J. (1979) Differential effects of two ways of devaluing the unconditioned stimulus after Pavlovian appetitive conditioning. *Journal of Experimental Psychology: Animal Behavior Processes* 5, 65–78.

MacPhail, E.M. (1998) *The Evolution of Consciousness.* Oxford University Press, Oxford.

Provenza, F.D. and Balph, D.F. (1987) Diet learning by domestic ruminants: theory, evidence and practical implications. *Applied Animal Behaviour Science* 18, 211–232.

Regolin, L., Vallortigara, G. and Zanforlin, M. (1995) Object and spatial representations in detour problems by chicks. *Animal Behaviour* 49, 195–199.

Reid, P. (1996) *Excel-erated Learning.* James & Kenneth Publ., Berkeley, California.

Roberts, W.A. (2000) *Principles of Animal Cognition.* McGraw-Hill, Boston, Massachusetts.

Rothenbuhler, N. (1964) Behavior genetics of nest cleaning in honey bees. 4. Responses of F1 and backcross generations to disease-killed brood. *American Zoologist* 4, 111–123.

Shettleworth, S. (2000) *Cognition, Evolution and Behavior.* Oxford University Press, Oxford.

Thorvaldsdottir, A.G., Provenza, F.D. and Balph, D.F. (1990) Social influences on conditioned food aversions in sheep. Ability of lambs to learn about novel foods while observing or participating with social models. *Applied Animal Behaviour Science* 25, 25–33.

5

Social and Reproductive Behaviour

Daniel M. Weary and David Fraser

What is Social Behaviour?

Social behaviour happens whenever two or more animals interact. Examples are all around us: dogs greeting one another, crows fighting over picnic scraps, geese flying in formation, a dairy cow licking her newborn calf, or even two people kissing at a bus stop. All are examples of social behaviour, although the form, development and evolutionary bases of these interactions are very different. In this chapter, we provide an introduction to the basics of social interactions among animals, describing the ways in which animals interact and the reasons why they do so.

Communication

Almost all social behaviour involves some form of communication. When a young calf becomes separated from the cow it will call repeatedly. When the cow hears these calls it will turn toward the sound, attempt to approach the caller, and vocalize. Thus, signallers can affect the behaviour of receivers by means of the signals they produce. Communication can occur through a range of modalities – by sound as in the current example, but also by smell, sight and touch (Hauser, 1996).

Acoustic signals can vary in their duration, pitch (vibration frequency), or amplitude. These features can be described and studied with the aid of spectrograms which give a visual representation of a sound. For example, a call of a cow separated from her calf is illustrated in Fig. 5.1. From this spectrogram we can see that the call is about a second long and consists of a series of harmonics, the loudest of which are from 100 to 1500 Hz. Relatively loud acoustic signals such as this call can be useful for communication over larger distances, but other calls are quiet and for use in close proximity, such as the contact calls produced by a cow and calf in the hours after birth.

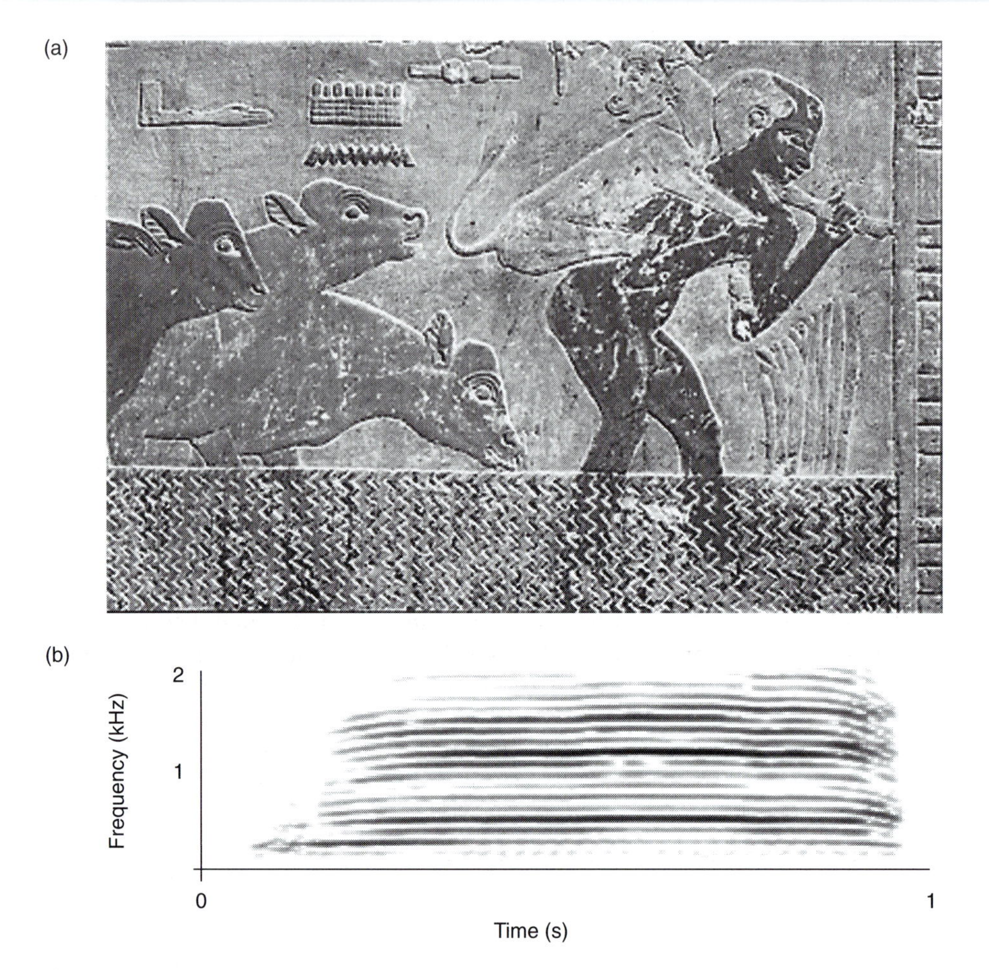

Fig. 5.1. (a) Separation distress in a cow and her calf, as captured by an unknown Egyptian artist (*c.* 2500 BC). (b) Modern methods of studying this and other types of social behaviour include spectrographic analysis. This figure shows a frequency spectrogram of a call produced by a cow soon after separation from its calf. Frequency spectrograms provide a graphical representation (frequency by time display) of a sound and facilitate the scientific description of the sounds that animals produce.

Humans, having a poorly developed sense of smell, find it hard to appreciate the importance of smell for many of the domestic mammals. Dogs, for example, are 100 million times more sensitive than humans to some chemicals. Many species use chemical signals to communicate such things as territorial boundaries and reproductive status. Scents have the advantage that they can be deposited and serve as markers long after the signaller has left. Some mammals also possess a vomero-nasal organ used to detect chemical signals; air is passed over this organ when animals show 'flehmen' or lip curling, such as when a stallion detects a mare in heat.

When animals are in relatively close range or live in an open habitat they can use visual signals, such as antlers on deer, colourful plumage on some birds, or body movements such as holding the tail erect by dominant dogs. Body movements have the advantage of being flexible and hence well suited for signalling information that is likely to change depending on the signaller's condition. More static displays, such as plumage coloration, are useful for signalling more stable information, such as an animal's species, sex and individual identity.

Tactile signals include social grooming in primates, suckling and nuzzling in mammals, and many of the behaviours associated with social play, such as biting and butting. One excellent example is the nuzzling of the sow's udder by her piglets during a nursing bout. Litters signal their desire to nurse by gathering at the udder and nuzzling it with their snouts, even though no milk is available at that time. In response to the nuzzling, the sow will often lie on her side, rotate her udder so that the teats are exposed, and grunt rhythmically. If enough piglets are present and continue nuzzling, the sow often responds with a brief milk ejection. Moreover, the sow signals the imminent availability of milk by an increase in her grunting rate. Thus the sow and piglets use both tactile and vocal communication to coordinate their nursing behaviour (this behaviour is further described in Chapter 11).

Regardless of how the communication occurs, we also need to understand how animals can benefit from producing signals, and how others can benefit from responding. For example, some animals vocalize in response to pain while other animals are stoic. This difference in behaviour probably reflects differences in the potential audience. A dependent young, such as a newborn piglet, may benefit by attracting a parent with its vocal response to pain. In contrast, an adult dairy cow may gain little from signalling pain associated with lameness. Indeed, in a natural setting, signalling pain might alert predators to the cow's higher vulnerability.

Living in Groups

Natural selection acts powerfully at the level of the individual. Thus a behaviour can be maintained by selection when it provides an advantage to the individual even if it has negative consequences for group mates. When a vulture joins other vultures at a carcass, it obviously benefits from access to food even if this means that other group members get less.

In other cases, known as 'mutualism', each individual can achieve some benefit from interacting. For example, foraging house sparrows may benefit from the arrival of other group members because this reduces their risk of predation or increases their foraging efficiency. In fact, a sparrow that discovers a rich food source will sometimes give 'food' calls that attract flock mates to the find.

In 'altruistic' interactions, an individual actually pays some cost by engaging in a social interaction that benefits another. The most obvious examples come from insects with sterile castes. Worker honeybees do not reproduce but instead spend their lives helping to rear the queen's offspring. This behaviour can be understood in part because of the benefits derived from kin selection: the workers are closely related to the queen and her offspring, and thus their behaviour helps to increase the frequency of the genes that they share with these relatives. The evolutionary biologist W.D. Hamilton devised a rule that could explain when this type of kin-selected altruism should occur. Hamilton stated that an individual should perform an altruistic act when the cost (C) of that act to the individual is less than the benefit (B) to the recipient divided by their coefficient of relatedness (r) (i.e. $C < B / r$). Thus identical twins ($r = 1$) may be expected to perform any behaviour when the benefit to the twin exceeds the cost to the performer. For full siblings ($r = 0.5$), we expect the behaviour only when the benefit to one sib is at least double the cost to the other. As the famous biologist J.B.S. Haldane explained, we should be willing to lay down our lives for two full siblings or eight first cousins!

Altruism may also occur between non-relatives if the individuals are interacting over a time scale that allows for reciprocation. Birds may benefit from flock membership if on different days different birds find food, thus allowing all members to achieve a more constant food supply.

Some of the most interesting instances of altruistic behaviour may be due to natural selection acting at the level of the group rather than the individual (Sober and Wilson, 1998). In these cases individuals pay some cost, such as reduced reproduction, to the benefit of the group. The conditions under which such 'group selection' can occur in nature are limited, but artificial selection in domestic animals can be arranged so as to favour altruistic behaviour among group members. If poultry geneticists select individual birds with the highest egg production in a group, they may be breeding inadvertently for aggressive, competitive behaviour. If, however, they select for whole cages of related birds that achieve high production on average, then they will breed for an ability to do well in a social setting. Given that laying hens are typically housed in groups on commercial egg farms, the latter would seem to be a better strategy. Indeed, experimental work has shown that such group-level selection can lead in just a few generations to birds that are both highly productive and relatively non-aggressive.

One of the most basic reasons for forming groups is safety in numbers. If all else is equal, a member of a pair should be half as likely to be eaten by a predator as would the same animal on its own. In this way animals may be able to dilute their risk of predation by forming simple aggregations (Treves, 2000), although this benefit will be reduced if larger groups are more likely to be detected. Risk dilution may also be shared unequally among group members. For example, animals may benefit more from group membership if they are at the centre of the herd rather than on the periphery, or by associating with more vulnerable herd mates. According to Canadian folk wisdom, it is best to travel in bear country with companions that you can outrun in a foot race.

Group membership may also help in detecting danger. In larger groups there are more eyes, ears and noses for early detection of predators. Thus groups may be able to detect danger more reliably or more quickly than solitary individuals. Also, frequent scanning for predators reduces the efficiency of other behaviours such as foraging. By forming an association with others that 'share' the cost of scanning, each group member may be able to forage more efficiently.

An obvious cost of foraging in a group is that animals may have to share the food that they discover (Giraldeau and Caraco, 2000). When foraging in a group, individuals can take on different strategies or combinations of strategies: some may be food finders while others are 'scroungers' that simply exploit the food. For example, a foraging bald eagle can look for salmon on its own, or simply wait for another eagle to find the fish and then steal or share in the meal. The choice of strategy will depend partly on the proportion of individuals using the two strategies (scroungers will do well when they are few) and any inherent advantage of being the finder (such as gaining longer access).

Group foraging can also have mutual benefits for group members. For example, when food sources are distributed in patches with relatively barren areas in between, foraging as a group can improve the detection of these patches. This benefit will reduce the variation in intake over a fixed period of time, a factor that may be particularly important to animals such as small birds that face the risk of starvation overnight. Feeding in groups can also improve the probability of capturing and consuming prey items, especially for animals capturing larger prey that can be difficult to subdue. For example, lions are more likely to capture zebras if hunting in pairs or larger groups than when hunting alone.

Competitive interactions

Why do some animals defend resources from competitors whereas others simply exploit resources without trying to defend them? Active defence of resources involves costs as well as benefits, and the net benefit of resource defence is only sometimes positive. For example, in so-called 'scramble competition', a pig can try to out-compete its pen mates simply by eating faster when food is scattered on a floor. Alternatively, a piglet like the one illustrated in Fig. 5.2 can attempt to exclude others from a food source by active 'territorial defence'. Which of these two patterns we witness will depend on how resources are distributed in both time and space, and by characteristics of the competitors.

Imagine a group of sows foraging for worms in a freshly ploughed field. The worms will be relatively evenly distributed in space so there is little to be gained from defending any specific part of the field – the time

Fig. 5.2. Piglets fighting at the udder. Fighting is common within the first few days after birth as piglets compete for access to a teat that they then defend throughout lactation (Photo by D. Fraser.)

spent in defence would only be lost from foraging. Now imagine this same group of sows fed a concentrated diet from a small feeder. In this situation defence of a specific area (the feeder) may well pay off, as time and effort spent excluding competitors may allow the territory holder to consume extra food. As a rule, it is less beneficial to defend resources spread over a large area. How resources are distributed over time will also affect how animals benefit from defending these resources. For example, feed delivered gradually over the day as a slow trickle is more easily defended against competitors than an equivalent quantity delivered all at once.

When resources are defended, the outcome of contests may be decided by differences in body size, strength or competitive abilities, factors collectively known as 'resource-holding potential' or RHP (Pusey and Packer, 1997). Animals that compete for resources need to become skilled at assessing each other's RHP so as to avoid the risk of injury and wasting effort from engaging in fights the animal is likely to lose. Newly mixed pigs spend a considerable time fighting, and these fights can lead to injuries. However, these fights are longer and more serious when animals are similar in RHP. Thus mixing animals that differ in body size, for example, will help to reduce the amount that these animals fight.

Sometimes other sorts of asymmetries can affect which animal wins a contest. One of the best known relates to the advantage of being the current resource holder. Territory holders may benefit from better knowledge of the territory. Also, intruders must often compete not only with the resident but also with the neighbouring territory holders – a cost that the resident does not have to pay. Animals can also vary in their need for a specific resource, and this too can affect the competitive behaviour. For example, a lioness may defend a portion of a carcass for a period of time, but as she becomes satiated she is more likely to be usurped by hungry pride mates.

When individual animals are in frequent competition over resources, they can avoid the costs associated with this competition by establishing which one is 'boss'. In the 1920s an ethologist, observing small flocks of chickens, noticed that one bird in each flock tended to peck all the others, whereas a second bird pecked all but the first, a third pecked all but the first two, and so on. This simple relationship came to be called the 'peck order' or 'dominance order' of the flock. This idea took wing, and soon animal behaviourists were describing social behaviour in other situations, from rat colonies to office politics, in terms of dominance orders, often by testing animals to determine which would have priority of access to resources such as food.

In reality, social relationships between animals are generally far more varied and complex than a simple dominance relationship would suggest. For many species, although it may be possible to rank the animals based on their access to resources, a simple hierarchy may misrepresent the kind of social behaviour seen in the group. In long established groups of pigs, for example, the animals often show mild mutual aggression with no clear hierarchy. Other species have multiple forms of social relationships: in cattle, for instance, individuals with high priority of access to food may not be the ones that normally lead the herd to new locations. A focus on dominance also tends to emphasize competition between animals, whereas some relationships are characterized more by cooperation and mutual tolerance. Pairs of chimpanzees in a group show forms of friendship and mutual support that are as varied as the personalities of the animals. Thus, while dominance over access to resources may be an important element of social behaviour in many species, attempts by scientists to determine dominance orders may reflect the scientist's preconceptions more than the true nature and range of social behaviour seen in the species.

Sexual interactions

All sexually reproducing species, including all birds and mammals, must at a minimum interact with one other individual in order to reproduce. One critical issue in reproduction is choosing a mate. The importance of mate choice to males and females depends on the time, energy and other resources that they invest in producing the offspring (Clutton-Brock, 1991). Some animals contribute only gametes (sperm or eggs) that may be relatively inexpensive to produce. In other instances pre-mating investment in gametes or post-mating investment in parental care may impose large costs. In general, we expect the sex that makes the larger investment (and thus has a lower potential reproductive rate) to be the one that exercises the most choice. In all of the common domestic animals, the male is able to father young at a faster rate than females can produce them, so we typically see males competing for access to females, and females being selective in their choice of mates.

Mating systems can be divided roughly into four categories: monogamy, polygyny, polyandry and promiscuity. In monogamous systems male and female bond for at least some period and often both parents contribute to care of the offspring. Polgynous systems are characterized by males mating with several females and females mating only with a single male and normally caring for the young. Polyandrous systems are the reverse. Promiscuous systems involve a mixture of polygyny and polyandry. Domestic animals typically have promiscuous or polygynous mating systems.

Polygynous males can gain mating opportunities by controlling access to resources important for female reproduction (e.g. food or nesting sites). This 'resource defence polygyny' will be facilitated by factors that help make resources defendable, such as the clumping of suitable resources in space. Males that control key resources will be more likely to mate or find multiple mates. For example, male red-winged blackbirds that have territories containing the most suitable nesting sites are more likely to attract multiple females as mates.

Alternatively, polygynous males can attempt to defend a group of females as a harem. This is facilitated when females group for other reasons, such as for anti-predation benefits as discussed above. Red deer stags, for example, compete with other males for control of harems, and a harem keeper must also spend time keeping the female group intact.

Parent–Offspring Interactions

Some of the most important social interactions occur between parents and their dependent young. Parents typically prepare suitable locations for the young, provide them with nutrients, and protect them from harm, although how these goals are achieved varies enormously from species to species.

Pigs and sheep provide an interesting contrast. The sow gives birth to a large number of small, fragile piglets. Typically she separates herself from her usual social group a day or so before farrowing, and seeks a secluded nest site. She prepares the site by rooting the soil to form a soft depression, and then lines it with resilient material such as branches, and with soft, insulating material such as grass. While the piglets are being born she lies relatively immobile with the udder exposed, leaving the piglets to find the teats unassisted. The carefully prepared nest site protects the piglets from cold and keeps the young in a single location which the mother can protect. With her relatively passive behaviour during parturition, the sow avoids harming the piglets by excessive movement, but does not learn to discriminate between her own and foreign young for perhaps a day or more. This interesting aspect of pig behaviour is further discussed in Chapter 11.

The ewe, normally giving birth to only one or two large, mobile offspring, follows a different pattern. She may move away from other sheep for parturition, or give birth in the midst of the flock. Once a lamb is born, the ewe licks it vigorously; this stimulates the lamb to rise and suckle, and it exposes the mother to the odour of the lamb, such that she can tell her lamb from others soon after birth. Although the lamb and ewe are highly mobile, they remain close together in the flock through mutual recognition and attraction. Further discussion on this subject can be found in Chapter 10.

Given its importance to fitness, parental behaviour is underlain by strong motivations. Under the influence of hormonal changes before parturition, the sow becomes extremely restless, and even if confined in a narrow stall will still greatly increase her level of activity, standing and lying repeatedly and showing a strong interest in nesting material. Once the young are born, the sow may become protective if intruders approach the nest, and if a piglet escapes from the nest and gives characteristic 'separation' calls, the sow becomes very attentive and seeks out the lost animal. Maternal motivation in the ewe takes a somewhat different form. Ewes that are about to lamb often become intensely attracted to newborn lambs, and may even 'steal' newborns from other ewes. Once her own is born, however, the ewe will often react aggressively to foreign young that try to suckle from her teats.

Traditionally, many animal behaviourists studied parent–offspring behaviour as if the relationship were all one-way, with the parent providing care for relatively passive young. In reality, the young have signals – sometimes subtle ones – that solicit care from the parent. Many young mammals and birds have some form of distress call when separated, and hunger signals which encourage the parents to bring more food. In a famous demonstration, an ethologist played 'begging' calls to parental swallows from a speaker concealed beneath their nest; the parent birds, being exposed to 'begging' calls from both the speaker and their own nestlings, brought unusually large amounts of food to the young.

Given the ability of young to stimulate parental care, parents need to strike a balance between providing too much care and too little. If parents devote too many resources to their present young, they may be slow to re-breed or not be in good enough condition to raise the next young successfully. The young, however, having a greater stake in their own success than in that of future siblings, may solicit a higher level of care than the parents ought to provide for their own maximum reproductive success. One obvious outcome of these conflicting interests is 'weaning conflict' whereby the parent may repel attempts by young to suckle or solicit food, or may even drive their young away at a certain stage in their development. In a related phenomenon, mothers (such as lionesses) that are capable of raising several young at once may, if offspring die and leave her with only a single cub, abandon the survivor and re-breed rather than continue a lactation for only one infant.

Play

One of the most interesting but under-researched aspects of social behaviour is play (Spinka *et al.*, 2001). Social play often consists of interactions that imitate the process, if not the end point, of more clearly functional behaviours, such as fighting. Sometimes animals use specific behaviours that signal the onset of play. Domestic dogs, for example, will often 'bow' during play sequences, especially if play includes fighting behaviours that could trigger aggression from the playmate.

How animals benefit from play is not well known, but play could help animals improve social skills, especially for species that need practice to develop effective courtship, appeasement or competitive behaviour. For example, pre-weaned dairy calves reared in groups (as compared to the more conventional individual housing) spend time playing, and these calves are more likely to become dominant when mixed with animals that have been individually reared. Play may also prepare animals for coping with unusual situations, such as maintaining balance on a slippery surface. Indeed, play sequences often involve some aspect of self-handicap, such as an ungainly body posture, which may make an animal particularly likely to fall or require it to move in unusual ways.

Human–Animal Interactions

For most animals in the wild, social behaviour involves principally interactions between members of the same species. For domestic animals, however, social interactions with humans can also be important. Humans are remarkable for forming close relationships with a wide variety of other species. When this happens, the human–animal relationship often builds on a predisposition of the animal to interact with members of its own species in characteristic ways. Dogs may redirect their natural protectiveness toward their human families, guarding the home and its inhabitants against intruders. Dogs may also form dominance relationships with humans, deferring to certain members of a family but not others.

Young birds and mammals tend to form strong attachments to their own mothers, and may, at the same stage in their lives, form similar attachments to human handlers. Hand-raised lambs are famous for forming close attachments to a human care-giver and following the person as they would follow their own mother. In many cases, such learning happens successfully only during certain 'sensitive periods'. Moose, if taken from the wild and raised by humans from birth, become so tractable that they can be hand-fed and used as pack animals throughout their lives. If, however, the moose spends a week or two with its mother, it may never become comfortable in the close presence of humans. Dogs appear to have a sensitive period for socialization with people at the age of roughly 2–3 months, perhaps corresponding to the age when puppies would normally meet dogs other than their own mother and siblings (see Chapter 12 for further discussion on the behavioural development of dogs). If socialization to people does not happen at that stage, the dog may never become satisfactorily responsive to humans. In some species, the early learned attachment ('imprinting') also affects later reproductive behaviour. Artificially reared chicks, ducklings and goslings that become imprinted on a human, and follow this person as they would follow their own mother, may then direct courtship toward humans later in life.

Animals can also develop learned fear of humans (Hemsworth and Coleman, 1998). In a famous study, Paul Hemsworth explored why nine small one-person pig farms, all using the same feed and same type of pigs, had very different levels of productivity. He found that on high-producing farms, the sows approached human visitors confidently, whereas on low-producing farms the sows tended to shy away. He concluded, through a series of experiments, that rough or unpredictable handling by the farm operator can lead to a learned fear of people, and that the stress caused by this fear can interfere with breeding and other elements of productive efficiency (Fig. 5.3). In general, a rapid and loud approach by a person standing tall above the animals tends to create a fear in animals, whereas a slow, quiet, crouched approach minimizes the animals' fear response.

Fig. 5.3. An example of inappropriate pre-slaughter handling of pigs, as depicted by J.F. Millet's *Killing of the hog* (*c.* 1870).

For millennia the domestication of animals has depended on the ability of humans to understand the social behaviour of animals, to accommodate this behaviour in controlled circumstances, and sometimes become part of the animal's social world. There is great scope for improving our care and handling of domestic animals by further improving our understanding of their social behaviour.

Most of the theories, concepts and vocabulary presented above were developed by behavioural scientists to help us understand how behaviour equips animals to live in the environment in which the species evolved. With most domestic animals, many of the behavioural adaptations of the species are still present, but may be ill-suited to the unnatural physical and social environment in which the animals are kept. Like its wild counterpart, the newly hatched domestic chick appears predisposed to become imprinted on a parental figure, but what form does imprinting take when hundreds of chicks are hatched together in an incubator? Like the wild sow, the domestic sow seems predisposed to wean her young gradually by spending less and less time with them as they age, but what are the consequences if the sow and litter are penned together continuously for several weeks, and then abruptly and permanently separated? Females of many species seem predisposed to select mates based on certain attributes, but if they are penned with only a single male, does this affect their willingness to mate or how strongly they display oestrus? One of the key challenges in the study of domestic animal behaviour is to understand how the rearing conditions are, or are not, suited to the behaviour of the animal. This understanding will allow us to design physical and social environments for animals that better accommodate their natural behaviour, thus helping to avoid problems for both the animals and the people that work with them.

References and Further Reading

Clutton-Brock, T.H. (1991) *The Evolution of Parental Care.* Princeton University Press, Princeton, New Jersey.

Giraldeau, L.-A. and Caraco, T. (2000) *Social Foraging Theory.* Princeton University Press, Princeton, New Jersey.

Hauser, M.D. (1996) *The Evolution of Communication.* MIT Press, Cambridge, Massachusetts.

Hemsworth, P.H. and Coleman, G.J. (1998) *Human–Livestock Interactions: the Stockperson and the Productivity and Welfare of Intensively Farmed Animals.* CAB International, Wallingford, UK.

Keeling, L. and Gonyou, H. (eds) (2001) *Social Behaviour in Farm Animals.* CAB International, Wallingford, UK.

Pusey, A.E. and Packer, C. (1997) The ecology of relationships. In: Krebs, J.R. and Davies, N.B. (eds) *Behavioural Ecology: an Evolutionary Approach.* Blackwell, Oxford, pp. 254–283.

Sober, E. and Wilson, D.S. (1998) *Unto Others: the Evolution and Psychology of Unselfish Behaviour.* Harvard University Press, Cambridge, Massachusetts.

Spinka, M., Newberry, R.C. and Bekoff, M. (2001) Mammalian play: training for the unexpected. *Quarterly Review of Biology* 76, 141–168.

Treves, A. (2000) Theory and methods in studies of vigilance and aggregation. *Animal Behaviour* 60, 711–722.

6 Behavioural Disturbances, Stress and Welfare

Linda Keeling and Per Jensen

When humans first started keeping animals in captivity, their concerns were probably restricted to how the animals could be prevented from escaping and how they could be kept alive and healthy. Later, we started to be concerned about production and how we could get our farm animals to produce more milk and eggs, grow faster and have more offspring or, in the case of our sport and companion animals, run fast and look more beautiful. During these times behavioural disturbances and stress were only a problem in so far as they affected health and performance, and good health was usually considered synonymous with good welfare. It was not until the 1960s and 1970s that we started to question these assumptions and to consider behaviour as an important component of welfare. This chapter deals with the important aspects of behavioural disorders and stress, and ends with a short review of current theories on what animal welfare is and how to measure it. Other books taking up this subject in more detail include Appleby and Hughes (1997) and Broom and Johnson (1993).

Behavioural Disorders

Normal and abnormal behaviour

One way to define what we mean by normal – or natural – behaviour of an animal could be to say that it is behaviour that has developed during evolutionary adaptation. This would include any learnt behaviour that serves the function of promoting the health, survival and reproduction of an animal in a certain environment. For domesticated animals, there are three important sources of information about normal behaviour. First, the behaviour of the wild ancestors; secondly, the behaviour of feral animals, i.e. domestic animals that have escaped or been released and have adapted to a life without dependence on humans; and thirdly, the behaviour of domestic animals when placed (usually by researchers) in environments similar to those of their ancestors.

Since learning and adaptation will modify the behaviour of any individual, there will always be a range of behavioural profiles that can be considered normal. Sometimes, people therefore dismiss the very concept of normal behaviour, and consider any deviations to be adaptive. However, in spite of large variation, there are aspects of the behaviour of any animal that are typical for the species (Fig. 6.1), and there may be aspects that are outside the range usually observed in members of the species in non-captive situations. It is therefore necessary to try to understand which behaviour patterns are species-typical and which ones can be regarded as not normal for an animal of a given species, sex and age.

It is perhaps a source of semantic confusion, but it is important to remember that, in this context, an abnormal behaviour may be very common among individuals of a species. For example, since most poultry in the world are kept in captivity, and most of the egg-laying hens are kept in cages, any abnormal behaviour caused by the cage environment is going to be shown by a majority of the hens. Frequency of occurrence, therefore, is no synonym to normality with respect to behaviour. When we consider abnormal behaviour and behavioural disturbances, we must always remember that the norm is the behaviour as it has evolved in the natural habitats of the species.

Fig. 6.1. When pigs are released into nature, they will form groups of closely related sows just like wild boars. During daytime they may spread out in pursuit of food, but during nights they will sleep close together in specially prepared night nests.

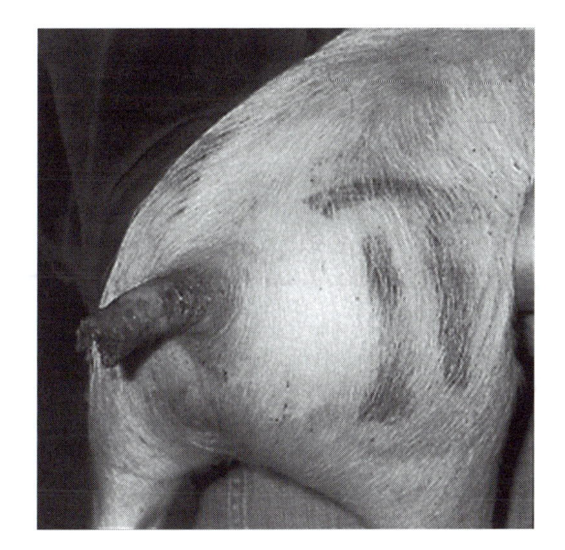

Fig. 6.3. Tail biting in fattening pigs is a serious welfare problem.

ponent, although in pigs given the opportunity to chew on blood-soaked 'imitation tails' (tassels of rope hung in the pen), nutritional deficiencies, and in particular salt deficiencies, were found to increase the incidence of chewing.

That there is some nutritional basis for cannibalism would be in accordance with outbreaks in the wild which often occur in times of food shortage. In rodents, pup-cannibalism is common. Here the cannibalism can be by the parent, which can sometimes be adaptive in nature. Often the victim would have had little chance of surviving in any case. However, it may also be caused by stress. When performed by an unrelated individual in nature, benefits of infanticide, i.e. eating young, may be increased food acquisition, but they can also be decreased resource competition or increased male reproductive success.

Despite the work on cannibalism from a behavioural ecology perspective, we have only a poor understanding of all aspects of cannibalism in farm animals. For the main part, it seems to be multifactorial, with factors such as bareness of the environment (e.g. lack of straw for pigs or litter for poultry, increasing stocking density and group size) all increasing the risk for cannibalism. There has been some success in reducing it by genetic selection, but whether this is a direct effect on cannibalistic behaviour or an indirect effect on some other linked trait is unclear.

Abnormal aggression

Aggression is used in the establishment of dominance relationships, and so in long-established groups the level is low and rarely a problem.

Aggression is therefore a normal behaviour, but in domestic animals we regard it as undesirable if it reaches proportions where it causes injury or practical problems. Unfortunately, the way we house or manage animals often results in high levels of aggression. For example, we often mix animals, or keep them in groups that would be unlikely to occur under natural conditions. As discussed in Chapter 8, horses live in bands in the wild. However, at a typical riding stable, horses may be moved from one paddock to another on an almost daily basis and may even be part-time group members, being outside only during the daytime in fine weather. Comparisons show that the level of aggression in these groups is much higher than in groups in the wild.

The most common reason for outbreaks of aggression is mixing of unfamiliar animals. This is frequently done, for example, with pigs. Attempts to prevent newly mixed animals from fighting have not been very successful, not even by using drugs.

Aggression is also influenced by stocking density, and an aggressive interaction may persist if a submissive individual is not able to signal its submission effectively because there is insufficient space for it to remove itself from the aggressor. This has shown to be a contributing factor in aggression between fattening pigs and sows.

Finally, aggressive problems in dogs are widely reported. Here the problem may be because of the high level of aggression, but it can equally well be that the owner perceives even a low level of aggression in their pet as undesirable, for example when there are small children in the family.

Stress

The concept of stress was developed over a number of decades in the middle of the 20th century. The leading persons were two physiologists, Walter B. Cannon and Hans Selye, who both studied bodily reaction patterns to potentially harmful stimuli. The reactions appeared to be more or less the same regardless of which types of threats they studied. Their combined findings gave rise to what we might call the standard stress model (Toates, 1995).

According to this standard model, a threatening stimulus (often referred to as a 'stressor') evokes mainly two sets of physiological responses (Fig. 6.4). One is the activation of the sympathetic branch of the autonomic nervous system, causing, for example, increased heart rate, increased blood pressure, decreased gastrointestinal activity, and increased secretion of catecholamines (adrenaline and noradrenaline) from the medulla of the adrenal cortex. Catecholamines, in turn, emphasize and prolong similar effects on the circulation and intestines. The other set of reactions consists of an increased secretion of the hormone ACTH (adrenocorticotrophic hormone) from the pituitary (regulated by CRH, corticotrophin releasing hormone) which stimulates

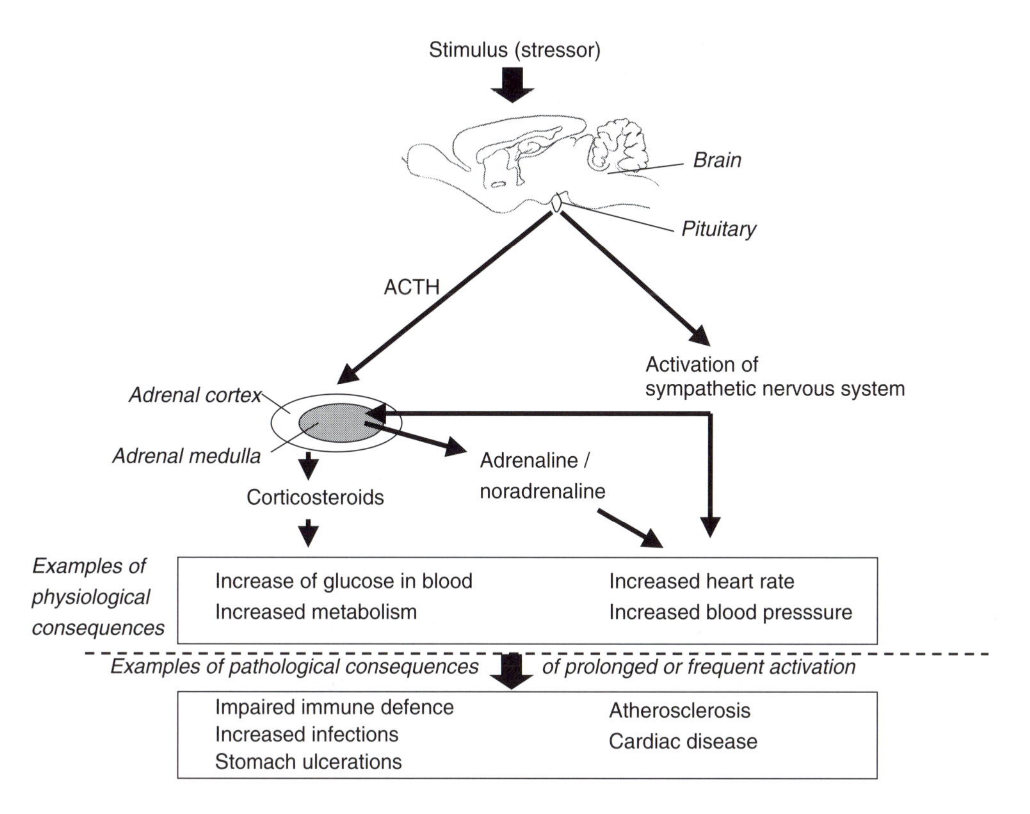

Fig. 6.4. The 'standard stress model', showing the two main physiological pathways: the hypothalamic–pituitary–adrenal cortex axis (HPA axis) and the sympathetic–adrenal medulla axis (SA axis).

secretion of corticosteroids (for example, cortisol and corticosterone) from the adrenal cortex. Corticosteroids are mainly metabolic hormones, stimulating the recruitment of energy in the form of glucose and fatty acids from deposits in the body.

Prolonged activation of this system was found to be harmful to the organism in a number of ways, causing syndromes such as stomach ulcerations, cardiovascular diseases and alterations in the efficiency of the immune defence. The systems have been found to be extremely complex and there are also several other systems involved, but basically, stress in this standard model is used as a concept describing the adverse bodily effects of certain stimuli.

Why should ethologists bother about this? There are a number of reasons. One is that the stimuli evoking the responses, the stressors, have often been described as 'psychological'. For example, a stressor could be to live in a crowded environment, or to be exposed to sudden and unexpected novelties, both of which are of interest to applied ethologists. Another reason is that the past few decades of stress research have shown that the standard model is not complete – stress responses are intimately connected with how animals perceive stressors, and to what extent they can act in a functional way in response to them. Hence, stress responses depend on the mechanisms controlling normal behaviour. Stress is therefore an important link between environment, behavioural disorders and disease (Broom and Johnson, 1993).

Predictability and controllability – key concepts in stress responses

The pathological consequences of a certain stressor have long been known to depend on how the stressor is perceived. The crucial experiment that demonstrated the salient factors in perception of threats was performed by Weiss (1971). He exposed rats to well-controlled stressors in the form of electrical shocks, delivered via an electrode attached to the tail. In each test, Weiss had three rats in single cages. Two of the rats were yoked in the same electric circuitry, so when one rat was shocked, the other received exactly the same sensation. The third rat was a control, which never received any shocks. One of the yoked rats was then given a predictive cue, in the form of a sound signal going off a few seconds before each shock.

The possibility of predicting the shock significantly decreased the development of stomach ulcers, even though both rats received exactly the same physical pain experience. When Weiss allowed one rat to exert some control over the shocks, the effect was even stronger. The rat could turn the current off once it had started (by turning a wheel), and thereby shorten the shock for itself and its unknown yoked partner, or postpone the shock somewhat. Again, ulcerations decreased dramatically, to the extent that the rats with possibilities to predict and control had only small differences in ulcerations compared to the animals which never received any shock at all.

This experiment, and others afterwards, have shown that the effects of a stressor do not depend so much on the physical characteristics of the stressor (intensity, duration, frequency, etc.) as on whether or not the animal can predict and, above all, control the stressor. Control is normally exerted by means of performing a behaviour which is relevant for the stimulus.

Individual differences in stress reactions

Even in historical stress research, it was clear that individuals often reacted quite differently to the same stressor under identical conditions. This obviously reflected some constitutional differences between individuals. In humans, psychologists would refer to 'Type A' persons – those more prone to react with activity in the sympathetic nervous system – and 'Type B' – those with less sympathetic reactivity. Behaviourally, Type A were described as more extrovert and aggressive, and more inclined to react with active attempts to control a stressor, while Type B would behave in much the opposite way.

This was paralleled by observations of similar differences in animals. Male tree shrews (*Tupaia belangeri*) were found to be either actively subordinate to a dominant male or passively submissive. If a submissive individual was placed in a cage close to a dominant male, it would lose weight and die within a rather short time, whereas the actively subordinate would survive and live fairly well in the same situation. In wild-caught mice, resident males (males which are the 'owners' of a cage) were found to behave in one of two distinct and different ways towards an intruder: either they attacked the new mouse within a very short time, or they did not attack at all. Very few intermediates, attacking for example after a few minutes, were observed, leading the researchers to conclude that there were two non-overlapping subpopulations among the original population of wild mice (Sluyter *et al.*, 1996). A strong genetic basis for the behavioural differences among the mice is shown by the fact that selection for either short or long attack latency in the resident–intruder set-up is possible. On average, the young of short-latency parents attack intruders immediately, and vice versa.

The mice differ not only in their response to strangers, but in more or less any situation where they are exposed to stressful events, even very minor ones. For example, mice of both lines take the same amount of time to learn to run down a maze to reach a goal box with food. But if a very small modification of the environment is introduced, such as placing a small piece of tape on the floor, the animals from the two populations differ in their reactions. Those selected for short attack latency seem not to pay any attention to the new item, whereas the long attack latency mice get completely disrupted in their task. They seem to forget what they are supposed to do and spend their time exploring the tape instead.

The reactions of short attack latency mice follow a general pattern, which has been termed proactive coping, whereas the pattern of the others has been called reactive coping. Proactive copers will attempt to deal with any challenge by finding routines and performing behaviour patterns that have proved to be successful earlier. Reactive copers face the same challenges by attempting to modify their behaviour to find the best way of handling every new potential stressor in its own way (Table 6.1). Physiologically, proactive copers react on challenges mainly with activation of the sympathetic nervous system, while this system is involved much less in the reactions of reactive copers.

It seems as if there is some generality to these observations across species and even phyla – there is some evidence of similar within-species differences among birds. In farm animals, the reactions of young piglets to a simple restraint test has been used to predict a variety of reactions to challenges later in life. It even has some predictive value for disease susceptibility and some aspects of meat quality after slaughter many months later (Ruis *et al.*, 2000).

Table 6.1. The major differences between proactive and reactive mice.

Character	Proactive copers	Reactive copers
Reaction to social intruder	High aggression	Low aggression
Time to learn a new task, e.g. running through a maze	Same as reactive	Same as proactive
Reactions to small changes in environment	Small, pays little attention	High, investigative
Time to adapt to major changes in environment, e.g. shift in diurnal light cycle	Long	Short
Activation of HPA axis in a stressful situation	Medium–high	Medium–high
Activation of SA axis in a stressful situation	High	Low
Typical pathological consequences of stress	Cardiac disease, stomach ulceration	Infection susceptibility, stomach ulceration

Is stress a physiological or a behavioural phenomenon?

Since stress was originally discovered by physiologists and described in physiological terms, it has often been treated mainly as a physiological phenomenon. However, the studies of Weiss (1971) and many other similar observations, have shown that whether a certain stressor is harmful or not to an animal depends on the animal's perception of the stressor and on its possibility to behave in a relevant way.

This has led some researchers to suggest that stress, in some aspects, can be regarded as mainly a behavioural phenomenon (Jensen and Toates, 1997). The common denominator of events that may become harmful to animals, according to this view, is that a behavioural system is motivated (Chapter 3), but the animal cannot perform the motivationally adequate behaviour, or at least, it cannot achieve the relevant functional consequence of the behaviour. The rats in the experiment of Weiss were exposed to electric shocks in the tail. This would have motivated their flight behaviour. By performing flight motions with the forelimbs, the rats acted with the relevant behaviour (so effectively 'controlling' the situation) and when the shock was turned off, they received the desired functional consequence. Consequently, those that did not achieve the desired consequence (the ones without control) were more harmed by the stressor.

In the example above, the animal is motivated to perform a behaviour by external stimuli (in that case, the electric shock), but they can also be motivated to perform a behaviour by mainly internal stimuli, such as hormones. For example, a sow due to farrow is motivated to build a nest, and this motivation is stimulated almost completely by internal hormonal events. If the sow is prevented from carrying out the nest-building activities, for example by being tethered in a crate, the increase in cortisol is much higher than if she is free to move around. By this measure, the sow would be said to be stressed because she is not able to act according to her motivational state.

Animal Welfare

The discussions about abnormal behaviour in domestic animals and improved knowledge of stress both contributed to a heightened awareness of animal welfare. But in reality it is not so simple, and not even correct, to think that good animal welfare (or animal well-being, as it is sometimes called) is merely the absence of behavioural disorders and stress.

What is animal welfare?

The Brambell Committee was the first to attempt a scientific definition of animal welfare in 1965. They were rather farsighted in three ways. First, they drew attention to the importance of behaviour in animal welfare. Up until that time good welfare had almost been synonymous with good health. Secondly, they stressed the importance of the scientific study of animal welfare, paving the way for future experimental studies. Thirdly, they accepted that animals had feelings, which went against the behaviourist trend of the time (see Chapter 1). The committee also proposed five freedoms that every animal should have, irrespective of how or why it is kept. These were that the animal could lie down, stand up, turn around, stretch and groom. While these may appear self-evident freedoms, there are many examples even today where they are not achievable, e.g. tethered cows and pigs which cannot turn around, or laying hens in cages which cannot flap their wings. The British Farm Animal Welfare Council later revised these five freedoms to the following:

- Freedom from thirst, hunger and malnutrition.
- Appropriate comfort and shelter.
- Prevention, or rapid diagnosis and treatment of injury and disease.
- Freedom to display most normal patterns of behaviour.
- Freedom from fear.

Since the Brambell Committee's first attempts, many others have attempted to define welfare, but, with hindsight, they can be classified into two main categories. One category emphasizes the biological functioning of the animal (its health, reproductive success, etc.), whereas the other emphasizes the subjective experiences of the animal (suffering, pleasure, etc.).

A well-accepted definition of the first type is proposed by Broom (summarized in Broom, 1996), who defines it by saying, 'the welfare of an animal is its state as regards its attempts to cope with its environment'. Here it is proposed that we can assess welfare by recording disease, injury, abnormal behavioural patterns and physiological changes related to stress (a high incidence of which would imply that the animal was not coping) and growth or reproduction (high rates of which would imply that the animal was coping). It is argued that by taking many measures from behaviour, physiology and health we get a balanced view of where the animal is on the scale from coping easily to not coping at all. A benefit of this approach is that we already have many techniques for measuring these parameters. The disadvantage is that we are not sure how to combine them in a way that takes into consideration the fact that some measures should be given more weight than others.

A well-accepted definition of the second type is proposed by Duncan (summarized in Duncan, 1996), who says that 'welfare is all to do with what the animal feels'. Here it is proposed that feelings have evolved in animals to improve survival and fitness. Pain is obviously adaptive; by making it unpleasant for an animal to put weight on an injured leg, the chances that the leg will heal increases. Similarly, the frantic behaviour seen in frustrating situations may increase the chances that the animal eventually changes its situation. It may even be that positive states, such as pleasure, are used for rewarding the animal for performing an appropriate behaviour, so increasing the chances that it is shown again.

It is argued that while under natural conditions feelings will closely reflect the health and physiology of an animal, there is a risk that for domesticated animals kept in captivity, the feeling and its usual physiological correlate become separated. In view of this risk, it is argued that we must pay attention to whether the animals feels hungry or feels stressed, not necessarily to whether the animal has a nutritional deficit or elevated corticosteroid levels. The advantage with this approach to defining animal welfare is that it is intuitively appealing, reflecting concerns in studies of quality of life in humans. The disadvantage is that as yet we do not have good techniques for measuring emotional states in animals. To have any indication of what an animal feels, we have to take indirect measures, mainly by examining their behaviour.

Assessing welfare

Whereas the two lines of scientific approach to welfare assessment may at first appear rather different, they differ mostly in where the emphasis is placed; the immediate subjective experiences of the animal or its long-term biological functioning. When it comes to the actual methods used in the scientific assessment of animal welfare, there is actually good consensus between researchers. In the next section we will go briefly through the health, production and physiology measures and then, in this book on ethology, spend most time on the behavioural indicators that have been used to assess welfare (Mason and Mendl, 1993)

Health and production as indicators of animal welfare

It is obvious that animal health is an important aspect of animal welfare. An animal with a broken leg, bleeding wound or other physical injury clearly has poorer welfare than an animal without that damage. But not all health issues are so clear, and it is important to remember that the border between health and disease is often indistinct. Complete health, that is to say 100% disease-free, probably doesn't even exist since the body is continually reacting to bacteria and viruses. Furthermore, even an unhealthy animal does not necessarily experience pain or distress. A dog with a tumour probably does not experience any pain or distress initially and a chicken with weak bones will not experience pain until there is a fracture. But the dog will probably suffer reduced welfare later, if the tumour is not treated successfully, and the hen is at risk of reduced welfare since bones are often broken during handling and transport.

A simple conclusion therefore is that an unhealthy animal either already has its welfare reduced or is at risk some time in the future of having its welfare reduced. But, as will be explained later, there is more to good welfare than being healthy.

Just as historically a healthy animal has been regarded as an animal with good welfare, so it has been argued that a farm animal with a high production has good welfare. But this view is criticized because production traits, e.g. number of offspring, growth rate, etc., have been the focus of intense selection for many generations. It differs between breeds and strains even under identical housing and management conditions, and it can be manipulated easily by changes in diet, lighting schedules and so on. In fact, very high production itself can lead to problems, such as udder inflammation in high-producing dairy cows or leg problems in fast-growing broiler meat chickens and fattening pigs. These have been called production diseases and are a relatively new category of welfare problem resulting from high production. A final, albeit extreme, example to demonstrate the risk of using traditional production parameters as welfare indicators, is the case of number of offspring. Using this measure, the animal with highest welfare would be a breeding bull at a semen station whose sperm are used to inseminate cows around the world. In fact, this bull may even be dead!

In summary, while poor production performance may be associated with poor welfare, good production *per se* is no guarantee of good welfare.

Physiological indicators of welfare

The most commonly used physiological indicators are those associated with the stress response outlined earlier in this chapter. Commonly used variables therefore include heart rate, corticosteroid levels and adrenal gland weights. More recently some of these measures have been taken remotely using radiotelemetry or by measuring corticosteroids in non-invasive ways, from the saliva, urine or faeces, instead of from blood samples. Not disturbing the animal while taking the measurement is particularly important when the physiological response is rapid, such as is the case with heart rate.

The frequently used interpretation is the greater the stress, the poorer the welfare. Usually this is true, but not always. The physiological response to short-term (acute) stress is different from that to long-term (chronic) stress because the system adapts and downregulates. Also some so-called stress responses can reflect positive experiences such as excitement and arousal.

To help overcome the problem of there being no obvious cut-off point where the level of stress can be said to be 'bad', some researchers have proposed that there is a welfare problem when the stress response is such that it leads to a change in the biological function of the animal and the animal enters a prepathological state. In short, it is argued that instead of measuring the stress response *per se*, the consequences of the stress are measured. Examples of such consequences can be immunosuppression or reduced reproductive success.

Behavioural indicators of welfare

The advantage of behavioural indicators of welfare is that they are the easiest to obtain and they probably reflect an animal's first attempts to cope with a less than optimal situation. Thus responses such as huddling together when the temperature is too low or vocalizing when hungry probably occur long before there is a threat to welfare from hypothermia or starvation. Natural selection has led to mechanisms that operate earlier and earlier, thus giving us more sensitive indicators that welfare is at risk than the obvious indicators such as injury and disease (Dawkins, 1998).

Comparison with normal behaviour

Early work on behaviour related to welfare was often associated with developing new housing systems and it was common practice to compare the time budget of animals in the new system with that in the traditional system (a time budget refers to how much time an animal allocates to different activities over the day). Comparisons were also made between animals kept under traditional systems and those kept under extensive conditions in outdoor enclosures. These extensive conditions were assumed to be better for welfare and the standard upon which other systems could be compared. But there are several problems with such comparisons. That the time budgets of animals differ according to the environment in which they are kept is not surprising. It is much more difficult to interpret this difference in terms of the animal's welfare. If the range of behaviour patterns seen in one system is less than in the other, it may mean that the behaviour is not released by that particular environment or it may mean that the animal is prevented from performing the behaviour.

An example of the first type may be anti-predator behaviour. Under extensive conditions poultry can give alarm vocalizations and escape to cover several times per day. If this behaviour is not seen under more confined conditions, we would not recommend that the caretaker takes steps to elicit it. The behaviour is triggered by external stimuli in the environment and there are unlikely to be welfare problems associated with its absence.

An example of the second type may be roosting behaviour, which is also in part an anti-predator behaviour, since predation losses are lower for birds roosting up off the ground. Given perches, birds will usually roost on them during the dark period, otherwise they roost on the floor. Roosting behaviour is triggered by external stimuli, in this case the decreasing light intensity. Birds have a dark period even in commercial poultry housing but there are no, or at least should not be, any real predators. The question is whether roosting up off the ground is still important for the birds and so whether perches should be provided under commercial conditions if welfare is to be maximized. We could not have answered this question without further studies which showed that indeed birds are motivated to get access to a perch for night-time roosting and so motivated for this behaviour. Thus comparisons of behaviour in different environments can often highlight issues to be investigated experimentally.

An additional benefit of studies comparing the behaviour or the same animals both in extensive and intensive conditions, or even of modern breeds and hybrids with their ancestors, is the important information we obtain on where, when and how different behaviours are performed. Such studies help us understand some of the behaviours we see in captivity but have not been able to interpret. For example, it is common knowledge that many birds build nests. It is therefore likely that the restless behaviour a laying hen performs in the 1–2 hours before she lays her egg may be related to nest building. However, who would propose a similar explanation for the movement patterns shown by a sow about to farrow, if there had not been observations on nest building behaviour in nature? After all, no other ungulates are known to build nests.

Experimental studies of behaviour as welfare indicators

Experimental studies on behaviour have contributed greatly to our understanding of animal welfare. Rather than observing the undisturbed animal, as described above, an artificial situation is created where the behaviour of the animal helps us answer questions about its state. Imagine that we did not know how a cow with a pain in its leg behaved. To determine this, we could inject uric acid into a joint (to create a form of gout which we know is painful) and in follow-up behaviour observations we would observe that the animal limped, putting less weight on the treated limb. In future, when we saw an animal limping in the same way, we might conclude that it, too, was experiencing pain.

Now we do not need to do this experiment, we already know the answer, but we can artificially create other states where we have less knowledge. Frustration is likely to result when an animal is prevented from performing a behaviour it is motivated to perform. By experimentally creating a frustrating situation, it is possible to study what behaviour patterns are shown by animals. Examples of frustration-related behaviour in chickens, for example, range from displacement preening when the level of frustration is mild, to increased aggression and stereotyped pacing in more frustrating situations. It is then possible to observe chickens in a variety of different situations to identify times of day or situations when there seems to be evidence of frustration, so that these situations can be studied in more detail. This idea of artificially creating emotional states has only been used to study frustration so far, but could probably be used to help investigate other states in animals.

Another methodology that has been used frequently is that of preference testing. Here the animal is given a choice between two or more resources or situations and the assumption is that it chooses in its best interest. The concept of preference testing is simple, but it is important that the tests are well controlled if the results are to be reliable. For example, many of our domesticated animals are wary of unfamiliar environments and food. Thus in a preference test the animal should have had equal experience of both choices. Animals also almost always choose to maximize short-term welfare over long-term, for example, a sweet tasty food over a nutritionally balanced one. Age, strain, previous experience and time of day will also influence the outcome.

Even with a well-designed preference test, though, there is rarely a 100% choice for one option over the other. Partial preferences may reflect that all animals rank the choices relatively similarly, or that a proportion of the animals prefer one option whereas the rest prefer the other option. Partial preferences may even reflect that one choice is preferred on one occasion and the other the next time, depending on the state of the animal. A modification is to leave the animal in the test apparatus for a longer time and then compare the total time spent in each of the available compartments. While this may overcome some difficulties with the simple choice test, one must remember that time spent using a resource does not necessarily reflect the whole picture of its importance to the animal.

Despite the improvements in preference test experimental design, one cannot compensate for the fact that in preference tests the choice is always relative. That is to say, while one option may be strongly preferred over the other, both may be poor for welfare or both may be good. Training the animal to perform an operant response such as pecking a key or pushing a lever to get access to a resource or an environment allow us to quantify the importance of a choice (Fig. 6.5). The assumptions are that if an animal will work for the resource then it must be of at least some importance to it and, presumably, the harder the animal will work, i.e. more pecks or pushes, the more important it is.

A further refinement of this idea of quantifying how hard the animal will work for the resource is to incorporate it into consumer demand theory used by economists to assess the importance of a commodity to human consumers. Commodities where demand decreases as price

Fig. 6.5. A laying hen pushes against a weighted door to open it and so gain access to whatever resource is provided on the other side. The weight of the door, and so the effort needed to open it, can be manipulated systematically.

increases are said to have an elastic demand, and those where the demand changes little as price increases are said to have inelastic demand. In humans, an example of the first may be fish and the second petrol. If the price of fish goes up people eat something else, but the majority of people still buy petrol for their car even if they complain about the price.

In animal studies, food is normally used as the standard example of an inelastic commodity and a regression line of data on amount of work to gain access to food with increasing price is usually rather horizontal. For other resources the regression line usually slopes downwards, showing that animals become less willing to obtain the resources as the amount of necessary work increases (Fig. 6.6). By comparing the slopes of the lines, one can rank the importance of different resources to the animals. Just as this technique can be used to measure strength of motivation to obtain a resource, it can also be used to measure motivation to avoid something unpleasant.

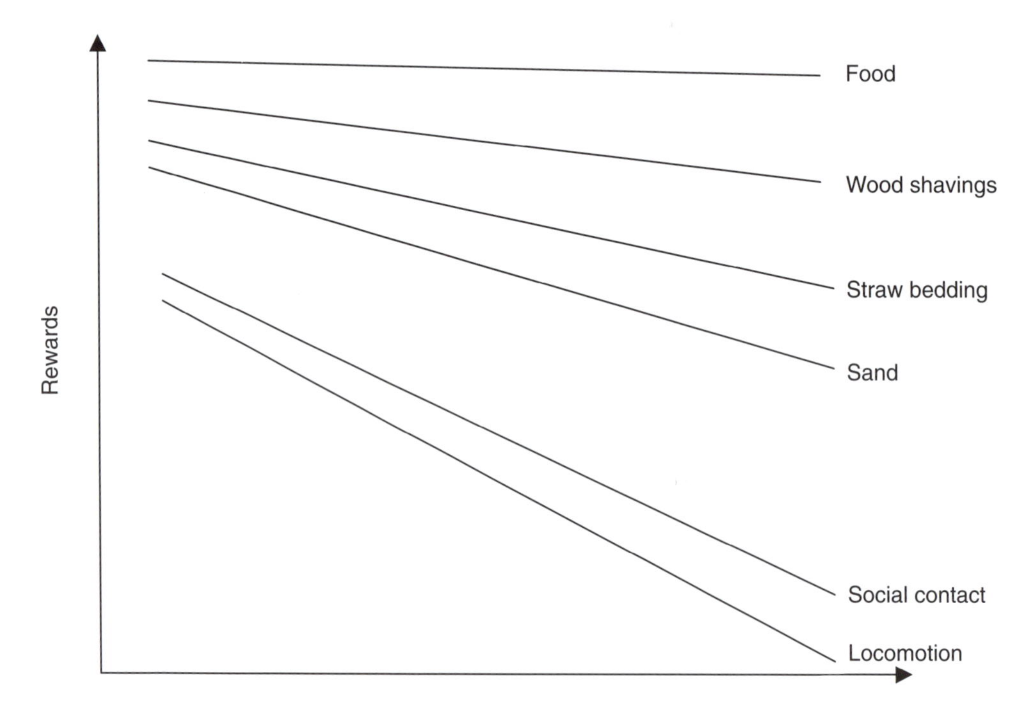

Fig. 6.6. Demand curves for different commodities in pigs. The pigs were required to perform an increased amount of work (numbers of snout presses on a bar) for access to a given commodity. The slope of the line indicates the relative importance of the commodity to the pig, a more horizontal slope indicating a higher importance. (Modified after Ladewig and Matthews, 1996.)

Concluding Remarks

This chapter has shown that scientific studies of animal behaviour can provide some of the necessary information for understanding the welfare of animals in captivity. Behaviour is often the first reaction used by an animal to adapt in a specific environment. Whereas data on physiological reactions and health are of course informative, they are often difficult to interpret in welfare terms without the necessary knowledge of accompanying behavioural reactions. Those will tell us whether, for example, an animal has an elevated corticosteroid level because there is an increased energy demand on the body, such as during play, or whether the animal is stressed and suffering increased disease risks. They will also inform us about whether the animal is in pain or otherwise suffering from a poor health state.

Of course, the ultimate decision as to whether a certain state of welfare is acceptable or not is an ethical one. It requires perspectives and arguments that cannot be obtained from scientific studies alone. However, any ethical discussion about what we should or should not do to animals should preferably be based on objective scientific knowledge about the state of the welfare of the animals.

References and Further Reading

Appleby, M.C. and Hughes, B.O. (eds) (1997) *Animal Welfare*. CAB International, Wallingford, UK.

Broom, D.M. (1996) Animal welfare defined in terms of attempts to cope with the environment. *Acta Agriculturae Scandinavica, Section A, Animal Science, Supplementus* 27, 22–28.

Broom, D.M. and Johnson, K.G. (1993) *Stress and Animal Welfare*. Chapman & Hall, London.

Dawkins, M.S. (1998) Evolution and animal welfare. *Quarterly Review of Biology* 73, 305–327.

Duncan, I.J.H. (1996) Animal welfare defined in terms of feelings. *Acta Agriculturae Scandinavica, Section A, Animal Science, Supplementus* 27, 28–36.

Jensen, P. and Toates, F.M. (1997) Stress as a state of motivational systems. *Applied Animal Behaviour Science* 54, 235–243.

Mason, G.J. (1991) Stereotypies: a critical review. *Animal Behaviour* 41, 1015–1037.

Mason, G. and Mendl, M. (1993) Why is there no simple way of measuring animal welfare? *Animal Welfare* 2, 301–319.

Moberg, G. and Mench, J.A. (eds) (2000) *The Biology of Animal Stress: Basic Principles and Implications for Animal Welfare*. CAB International, Wallingford, UK.

Ladewig, J. and Matthews, L.R. (1996) The role of operant conditioning in animal welfare research. *Acta Agriculturae Scandinavica, Section A, Animal Science, Supplementus* 27, 64–68.

Ruis, M.A.W., te Brake, J.H.A., van de Burgwal, J.A., de Jong, I., Blokhuis, H.J. and Koolhaas, J.M. (2000) Personalities in female domesticated pigs: behavioural and physiological indications. *Applied Animal Behaviour Science* 66, 31–47.

Sluyter, F., van Oortsmerssen, G.A., de Ruiter, A.J.H. and Koolhaas, J.M. (1996) Aggression in wild house mice: current state of affairs. *Behavioural Genetics* 26, 489–496.

Toates, F. (1995) *Stress – Conceptual and Biological Aspects.* John Wiley & Sons, Chichester, UK.

Weiss, J.M. (1971) Effects of coping behaviour in different warning signal conditions on stress pathology in rats. *Journal of Comparative and Physiological Psychology* 77, 1–13.

Part B Species-specific behaviour of some important domestic animals

Editor's Introduction

For the second part of the book, we turn our attention to the actual behaviour seen in some of the more common and economically important domestic animals. On the basis of the theoretical approach in the first part, it is now possible to go into detail as to what animals actually do.

The species dealt with in this book have been selected because they are the most commonly occurring of the domestic animals, and also those on which most research has been performed. For most of the species we will meet, there is considerable knowledge about the behaviour of their ancestors, and about the ways in which these animals react and behave under different husbandry conditions.

For each species, the authors have attempted to provide information on a number of different aspects. The origin and domestication history is given some emphasis in all chapters. Sections on social behaviour and communication, foraging and feeding habits, mating behaviour and other aspects of reproduction are then dealt with, although the exact order and content may vary somewhat between chapters, depending on the relevance for the species treated. Behavioural ontogeny is a further subject that is treated as far as possible, according to available information.

Most of the text is devoted to the normal behaviour of the species, but each chapter ends with a section on applied problems. This section covers typical issues raised by housing and husbandry conditions, including common behavioural disorders. Where information is available, the authors have attempted to provide some ideas about possible prevention or treatment of the types of problems described.

With the information provided in this second part of the book, it is hoped that students of animal behaviour will be better equipped to differentiate between normal and abnormal behaviour when animals are observed under practical conditions.

7 Behaviour of Fowl and Other Domesticated Birds

Linda Keeling

Origin and Domestication History

Many species of fowl have been domesticated, but by far the most common is the chicken (*Gallus gallus domesticus*). Both our commercial laying hens (layers) and our meat chickens (broilers) are derived from the jungle fowl, probably the red jungle fowl (*Gallus gallus*), which is still found in the shrub and bush of South-east Asia. It was domesticated about 8000 years ago for ceremonial purposes, because of its beautiful plumage, and for cockfighting. The Romans had a well-developed poultry industry, with breeds selected for high egg production. But following the decline of the Roman Empire, egg production did not again reach a commercial scale until the 19th century, when selection for breeds of birds to specialize in either egg production or meat production started in earnest. Egg-laying strains can be further divided into light hybrids, which are mainly white and derived from the White Leghorn breed, and medium hybrids, which are usually brown and derived from the Rhode Island Red breed, although the industry is continually selecting to reduce the body weight of the brown birds, so the difference now is not as great as it was some decades ago. More recently there has also been selection in broiler lines to produce a slowly growing bird for specialized markets as well as the more standard fast-growing broiler chicken.

Turkeys (*Meleagris gallopavo*), ducks (*Anas platyrrhynchos*) and geese (*Anser anser*) are also domesticated, as are quail (*Coturnix coturnix* and *Colinus virginianus*) and guinea fowl (*Numida meleagris*). Turkeys come from the New World and the most common commercial breeds are derived from crossings between the wild turkey of North America and turkeys that were originally taken to Europe by explorers visiting Central America. Although turkeys were probably already domesticated in Mexico in prehistoric times, they are a relatively recent addition to modern agriculture, having been kept for only a few centuries. There are many different breeds of ducks and geese, which have contributed to the breeds that are used today. They are now mainly used for meat production, although there are some very prolific egg-laying breeds of duck. All breeds of duck, with the exception of the muscovy, which is a South American species of duck, are descended from the mallard. Domestic geese are descended from the greylag goose and have been kept as guard animals as well as for food, but goose production has never reached the large-scale proportions of duck production. Even practices that have traditionally used geese, such as foie gras production, are now changing to use ducks. The Japanese quail is used mainly for egg production, while the bobwhite quail is kept mainly for meat production. Guinea fowl are widespread in Africa for both meat and eggs. Pigeons (*Columba livia*) are descended from the rock dove and, although domesticated some 500 years ago for meat, are now mainly kept for sport (racing and homing pigeons) or for pleasure (fancy breeds). Small birds, such as budgerigars and canaries have been bred in captivity for many years and are popular as easy-to-care-for pets. They have been selected for their colour and singing abilities.

Although many of the points taken up in this chapter relate to all birds, I will focus mainly on the most commonly occurring domesticated species, namely chickens and turkeys. This is because, commercially, they are the most important, and most applied ethology research has consequently focused on them. It is pleasing to report, though, that this is changing, and there is now much interesting applied ethology research on the other bird species.

Domestication is described more fully in Chapter 2, so it is sufficient to say here that in keeping with the majority of other species in this book, we have changed little in the behaviour of the birds we have domesticated, even if we have dramatically changed their appearance and their production. In laying hens this has meant that instead of laying one clutch of 8–10 eggs and then incubating them, as the jungle fowl does, the commercial laying hen goes on to lay a second clutch of eggs, and a third, usually without a pause between. Modern strains often lay over 300 eggs in 1 year. The commercial broiler chicken weighs 2.5 kg, over three times heavier than the adult jungle fowl, and it reaches this weight in 42–45 days. Turkeys, geese, ducks and bobwhite quail are all considerably heavier than their ancestors, reflecting the successful selection for meat production. In fact, selection for turkey breast meat has been so successful that these birds have a conformation that means they can no longer mate naturally and artificial insemination is normally practised.

Generally speaking there are two main categories of housing which, until recently, could have been divided almost exclusively according to the reason the birds were kept (Fig. 7.1). Egg-producing birds are generally kept in groups of 3–10 birds in cages, arranged several tiers high in long rows or batteries. Meat-producing birds (broilers, turkeys and ducks) are usually housed in large flocks of 2000–20,000 birds, with wood shavings or straw on the floor. However, in past decades, public criticism of cages for laying hens, especially in Europe, has resulted in an increase in loose housing systems for egg-laying birds also. These can be similar to floor systems for meat birds, but more often they have

(a)

(b)

(c)

Fig. 7.1. The most common housing system for egg-laying hens is the battery cage (a), although free-range egg production (b) is popular in some countries. Broiler meat chickens are usually kept in litter floor systems (c).

perches or raised platforms in them and so are called aviary systems. A good review of the different housing systems for poultry is given in *Poultry Production Systems* by Appleby *et al.* (1992).

Whatever the system, cage or loose housing, the degree of automation is usually very high. Food and water are delivered and eggs are collected automatically. While at first this seems to leave little scope for birds to express their natural behaviour, most behaviour patterns are still clearly evident in commercial housing systems. Behaviours such as feeding and drinking are obviously necessary for survival, and ground pecking and preening do not require any special additional resources. Other characteristics or behaviour patterns, such as precocial young which can be reared in the absence of a mother hen and egg laying in nest boxes from which the eggs can easily be collected, have been exploited by producers to make management easier. Nevertheless, as we start to understand more about bird behaviour, there is increasing pressure to allow those behaviour patterns thought to be important to the bird for its welfare, such as dustbathing and perching, even if their performance is neither necessary for survival under commercial conditions, nor directly advantageous to the farmer.

As outlined in Chapter 2, there are many characteristics that make it more (or less) likely for a particular species to be domesticated. Poultry and most of our other bird species have these prerequisites, which is probably why there are so many different types of domesticated bird. Among the most important of the prerequisites is social behaviour.

Social Behaviour

With the exception of geese and some quail, which are mainly monogamous, almost all breeds of domesticated fowl are promiscuous, with males mating with several females. In some species, such as the chicken, the male may defend his harem of females from other males. This mixed sex, mixed age group of birds has a well-developed social system and, in fact, it was while watching a flock of backyard hens that the concept of dominance hierarchies was first derived by the Norwegian researcher Shjelderupp-Ebbe. He called this the ‘peck order’, but it has since been seen in most social species. Dominance hierarchies, as they are more commonly called, have been discussed in more detail in Chapter 5, but it is worthwhile here describing the different forms of aggression that are seen in birds (see also Rushen, 1982).

Aggression in poultry can take the form of subtle threats and avoidances, pecks, and even fights and chases. Although the more severe forms of aggression are rare in stable groups of birds, they are more common when males or unfamiliar birds meet. During a fight, a bird leaps at the other bird, holding forward the spurs on the back of its legs. The beak is also a dangerous weapon and pecks from one bird to the comb of another often leave small wounds and scratches. Fights consist of repeated pecks, but usually pecks occur singularly and even one peck can be sufficient to establish dominance. Pecks are nearly always directed at the head of the other bird and are made up of a hard downward stabbing movement. Since threats take the form of intention movements to higher aggression, a bird gives a threat when it raises its head as though to peck another bird. Submissive gestures are when a bird lowers its head or turns away. Sometimes these gestures are so subtle that it is difficult for the observer to see that there has been a threat or a submissive signal given. Males and females usually have separate hierarchies and young birds are almost always subordinate to adults.

The natural group of jungle fowl consists of perhaps 5–30 individuals; a dominant male, females and their young. Juvenile males are subordinate to the dominant male and may even be expelled from the group when they mature. A flock moves within its home range, which may overlap with the home ranges of other flocks, but each flock returns to a specific roosting site as dusk falls. It is difficult to observe jungle fowl in their natural habitat since they are so timid, but there have been several studies of groups of jungle fowl in zoos (Dawkins, 1989). Turkey social organization may be similar to that of the jungle fowl, but they can also be found living in single-sex groups. During the breeding season males compete by giving displays for females at communal display grounds called leks. Each species has evolved to live in a social group that maximizes its chances of survival in its natural habitat, but in captivity we often keep them under very different conditions.

A major difference in the way we keep birds commercially is that they are almost always housed with individuals of the same age. This is to reduce the risk of disease being transmitted from older birds to younger birds, and it allows the poultry house to be thoroughly cleaned between batches of birds. In egg production, another major difference is that flocks consist of only females, and even in broiler production females may be kept separately from males, although these birds are usually slaughtered before they reach sexual maturity. Only in breeding flocks are mature males and females kept together.

These differences to the natural group composition can have consequences for the social organization of the flock. For example, it is known that the presence of males reduces aggression between females, and that this is the case even in large flocks, with only a few males. It is also known that in breeding flocks some males do not mate, because of the presence of other higher-ranking males, and that this can contribute to low fertility of the eggs from such a flock. Recently, there has been research on what a very large group size means for the formation of the dominance hierarchy. Originally it was thought that there would be problems with aggression in floor-housed and aviary flocks, because it would not be possible for a bird to recognize all the other individuals. But aggression in large groups is not especially high, and it has been proposed that there is no hierarchy in larger groups of birds because the costs of establishing and maintaining it outweigh the benefits in terms of priority of access to resources. Instead, it is proposed that birds use direct assessment of status based on cues such as comb and body size when they meet. More detailed information on poultry social behaviour is available in Mench and Keeling (2001).

Signals and Signalling Systems

As in all social species, birds have a well-developed system of communication. Of the many different modalities, visual and acoustic signals are the most important. Fowl have only about 26° of binocular vision, but good all-round monocular vision, which is typical for a prey species. They have good visual acuity, although it is not quite the same as ours, being based on four photoreactive pigments, instead of the three found in humans. They also possess coloured oil droplets in their cone cells, which filter light before it reaches the photoreactive pigments, and they are able to perceive ultraviolet light. As a result, birds may have good colour vision but perhaps require relatively bright light for this to function optimally. Individual recognition occurs mainly by visual cues, and it has been shown in studies where the physical characteristics of birds have been manipulated that it is the features of the head, such as comb shape, colour and size, that are most important.

Bird species are well known for their sometimes spectacular courtship displays. Our domesticated birds are not quite in the same league as some of the wild species, but visual displays are nevertheless important even in our most common breeds. The male turkey, in particular, has a colourful strutting display (Fig. 7.2), where the otherwise pale naked carunculated skin on the head and neck changes colour to become red and then blue. Indeed, most birds have some special display movements that are performed for the female in courtship or other males during aggressive encounters. Visual cues such as body and comb size are used for assessing dominance, and body posture is an important signal.

The vocalizations of domesticated birds are not well documented and we are far from knowing the function of all calls. It is clear that communication between the chick and the mother starts before the chick hatches, and in artificial incubators chicks give peeping calls. It is thought that this communication helps synchronize the hatching of the clutch and can modify the behaviour of the mother hen to increase or decrease brooding behaviour and so help with temperature regulation. Once hatched, chicks can distinguish their mother from other hens by her calls.

The laying call in hens, crowing in males and predator alarm calls are the vocalizations most easily recognizable in fowl. Audio-spectrographs of the different calls are given in Wood-Gush (1989). The laying call is also called the gakeln call and consists of an elongated note that increases slightly in pitch, followed by a succession of shorter notes. It possibly functions to attract the male to escort the hen to and from the

Fig. 7.2. Mature turkey stags have a distinctive courtship display. Unlike most other domesticated breeding birds, the two sexes are kept separately. The males are milked to collect semen, which is then used to inseminate the females.

nest, although it is also thought to be an indicator of frustration. The crowing of the cockerel is usually given in the early morning and is associated with territorial defence. The frequency of crowing is related to comb size, with larger, more dominant males crowing at a higher rate. Crowing rate is probably used by females and rival males as an indicator of fitness, since it is only larger, healthy males that can perform crowing at a high rate (Leonard and Horne, 1995). Alarm calls can be divided into aerial and ground predator alarm calls and are functionally referential. That is to say, they contain sufficient information for other birds to act appropriately. Ground predator alarm calls are harsh, with a quick start to the call, and elicit standing erect with a vigilant posture, whereas aerial predator calls gradually increase in intensity and elicit crouching or running for cover. Studies investigating audience effects on alarm calling have shown that males give most alarm calls when there is a female of the same species nearby.

Foraging and Feeding

Most domesticated birds are omnivores, eating both seeds and small invertebrates. They obtain their food by scratching and pecking the ground. Under natural conditions, chickens can spend over 90% of their time during the day in these activities. For ducks and geese, foraging can also involve filtering water and grazing. As in most species, feeding can be divided into appetitive, which is the food-searching phase, and consummatory behaviour, which is the actual eating of the food, and the proportions of these vary according to the abundance of the food and the social conditions.

Birds under commercial conditions are usually given free access to food in food troughs, but they still spend a large proportion of time ground pecking and scratching. Such behaviour is important for information gathering and monitoring the environment. Animals will even perform an operant task, such as pecking at a disk to get access to a food reward, even when there is food freely available in the pen. Working for food when it is freely available elsewhere is called contra-freeloading and is most likely explained by an individual's need to gather and update information about potential food sources. The only domestic fowl that are not given free access to food under commercial conditions are broiler breeders (Savory and Maros, 1993). These are the parent birds to the fast-growing meat (broiler) chickens and are kept on a restricted diet to minimize health and production problems that would occur if they were allowed to increase in weight to their full genetic potential.

Birds can be very choosy feeders and readily select the larger or tastier particles from commercial poultry feed. In commercial poultry houses food is often distributed around the house by a chain in the bottom of the feed trough. If this chain moves slowly, so that birds can peck at the feed as it goes by, the composition of the food changes as it goes around the house, resulting in the birds at the end sometimes receiving a less-balanced diet.

Birds have specific appetites for certain nutrients such as sodium and calcium, the latter of which is, of course, especially important in shell formation. Experiments have even shown that when given a choice of grain and a concentrate, birds can select the combination that is most appropriate for their needs, and this is sometimes used in organic production, when combinations of homegrown ingredients are provided.

Young chickens are precocial and will readily explore their environment, pecking at many potential food items, on the first day of life. While the tendency to peck is innate, chicks have to learn which substances are actually food items. They can do this by trial and error, or the mother hen can help by giving the tidbitting call and attracting the chicks over to where she is feeding. The food call of the broody hen also indicates food quality. Although chicks have the remains of the yolk sack in their abdomen, this is usually only sufficient nourishment for the first 2–3 days of life and so it is important that they learn quickly to identify food. Mortality because a chick failed to learn to feed is low in commercially reared chickens, but in turkeys, starve-out can be a problem.

As well as being more likely to peck at small, round items, since these have the greatest probability of being potential food, chicks will also peck readily at shiny objects. In nature this might increase the chance that the chick pecks at water droplets on vegetation or puddles on the ground. When drinking from the ground, a bird has to lift up its head between each mouthful so that the water can run down the oesophagus by gravity, although there is some swallowing. Birds usually alternate feeding and drinking over the day.

Biological Rhythms

Most animals are sensitive to increasing and decreasing daylength that, in nature, signals spring and autumn. Under commercial conditions, young females of egg-laying strains are kept on a short daylength and as the time for them to start laying eggs approaches, this is gradually increased to a maximum daylength (typically 16 hours) where it remains for the whole of the laying period. A decrease in daylength, especially as the birds become older, would result in the birds stopping egg production. Meat birds, on the other hand, are usually kept on a consistently long daylength, often 20–23 hours, so that there is the maximum time for them to feed.

Birds also show a relatively standardized diurnal pattern of behaviour. This starts with feeding behaviour and, if the bird is an adult laying hen, it usually lays an egg in the morning. This is described later in this chapter in the section on egg-laying behaviour. In the middle of the day, there is a peak in dustbathing (Fig. 7.3) and in the afternoon preening behaviour. During preening, birds rejoin the barbs on the vanes of damaged feathers using their beak. They also distribute oil over the feathers from the uropygial gland, which is located on the back at the base of the tail. This helps keep the feathers supple and in good condition.

Fig. 7.3. Dustbathing behaviour is shown in the middle of the day and is usually synchronized.

Dustbathing is performed once every 2–3 days and consists of birds lying and rubbing litter material though their feathers (Vestergaard, 1982). These rubbing movements are interspersed with bouts of wing tossing, a behaviour that is unique to dustbathing. The bird sits and uses its wings in upward movements to toss litter on to its back. This cycle of rubbing followed by tossing is repeated several times before the bird stands up and shakes the loose litter material from its feathers. In this form of 'dry shampoo' any excess lipids, as well as ectoparasites, are removed from the feathers.

Birds may rest in the afternoon if food is freely and easily available, otherwise there is more foraging behaviour. Mating behaviour in flocks usually occurs in the late afternoon. Finally, there is a second peak in feeding behaviour before the birds go to roost in the evening. Some birds are ground roosting, but the majority of the domesticated species roost up off the ground in bushes and trees whenever possible. Birds usually start to move towards the roost and jump up to it shortly before darkness. There is some evidence that birds have preferred places to roost and operant studies have shown that birds are motivated to get access to a perch for night-time roosting. Under natural conditions roosting up off the ground probably has survival value by reducing predation from night-hunting ground predators.

Mating Behaviour

In the natural situation, mating behaviour includes the defending of territories and the establishment of a harem or pair-bond, as well as actual courtship and mating. This is further discussed in Chapter 5, and here I will restrict myself to courtship and mating behaviour.

In chickens, the male can show one or all of a series of different movement patterns and vocalizations during courtship and mating. Tidbitting, although most often shown by a mother hen towards her chicks, can also be given by males to attract females away from the other individuals. The more obvious movements are waltzing, when the male circles the female with the outer wing trailing on the ground, and wing flapping, where the male flaps both wings so that they meet in the air above his back. Mating proceeds to mounting when the female crouches and the male steps on to the back of the female and there is cloacal contact, although forced copulations are not uncommon in certain strains. On some occasions females may solicit mating by approaching the male, and in some cases the male may dispense with all courtship and merely approaches the female and mounts. There has been considerable work on mate selection in females and it is shown that females prefer males with characteristics that indicate high fitness, such as symmetrical features or secondary sexual characteristics, such as spurs or decorative feathers, that are large compared with those of other males.

In commercial turkey production, sperm is collected from turkey stags and females are inseminated, since natural mating is no longer possible in broad-breasted breeds. Courtship behaviour in turkeys is characterized by slow movements and an elaborate feather display, with the tail being elevated and fanned and the wings lowered. The head is drawn in and the snood, a worm-like pedant just above the beak becomes elongated and the skin on the head and neck becomes coloured. The females show a very marked squatting behaviour.

Female birds can store sperm for up to 2 weeks, so it is not necessary for them to mate every day for every egg to be fertile. If a female mates with two males, it is usually the sperm from the most recent mating that fertilizes the egg, although it has been proposed that females can eject the sperm of lower-ranking males, so maximizing the chances that her offspring are sired by more dominant males.

Egg Laying, Incubating, Hatching and Parental Care

A hen usually lays eggs in a clutch, at which point she stops laying and starts incubating them. Modern laying strains, however, as a consequence of their selection for high egg production, do not usually show the hormonal changes that trigger the switch from egg laying to incubating, and so start to lay a new clutch of eggs directly after the previous one. Selection for egg production in meat strains has not been as intensive and so egg production in these lines is less impressive and birds do sometimes stop laying eggs and start brooding. Turkeys, in particular, are prone to broodiness. Some breeds have sex-linked traits, e.g. male and female chicks have different coloured feathers when they hatch or have different feather development, so that they can be sexed at 1 day old and reared in single-sex groups. This development of auto-sexing breeds has meant that selection can proceed separately for egg production and for meat.

Pre-egglaying behaviour, however, has not changed in any noticeable way, even as a consequence of selection, and hens still show the same phases of nesting behaviour as their ancestors and as birds kept in the wild (Duncan *et al.*, 1978). The first phase of nesting is nest-site selection. The bird may walk considerable distances during this phase and explore many potential sites before selecting one where she sits and starts to build a nest. Under commercial conditions the bird typically has her head held high and the front part of the breast is raised. The most important criterion for nest-site selection seems to be that the location is enclosed. In nature this may be under a stone or in thick vegetation, and in captivity it is hopefully in one of the nest boxes. Birds that are not provided with a nest box, e.g. commercial egg layers kept in cages, can spend longer in this nest-searching phase and may show signs of frustration, such as stereotyped pacing.

During nest building the bird scrapes out a hollow with her feet and rotates in the nest, defining the hollow even more with the front of her keel bone. She also rakes loose material towards the edge of the nest to build up a raised edge. The hollow shape makes it less likely that the eggs will roll out of the nest. Sometimes the hen will place long pieces of vegetation on her back, which is thought to break up her outline from aerial predators.

The actual egg laying is rather quick, with the bird taking a squatting position, often called the penguin position and expelling the egg. Although the bird may remain in the nest for a short time afterwards, she usually returns to the flock and starts to feed. When laying a clutch of eggs the hen should obviously lay all her eggs in the same nest. Under commercial conditions, where there are many potential nest sites, it is thought that nest boxes are super-normal stimuli and that birds have difficulty distinguishing between one nest and another. Confronted with many nest boxes, laying hens tend to choose those in the upper tier and those at the end of the row.

The whole process of selection and egg laying takes 1–2 hours in commercial laying hens and is under the control of the hormones oestrogen and progesterone. These hormones are released from the follicle after ovulation. It takes about 25 hours from ovulation until the egg is laid.

While brooding, birds leave the nest mainly to drink and defecate. They rarely eat and there is some evidence that they reduce their heart rate and probably also their metabolism. It has been estimated that a broody hen consumes only 20% of its normal daily food intake during brooding and loses 4–20% of its body weight during the 21-day incubation period. Once the chicks hatch, the hen and chicks leave the nest to forage.

For the first 3 weeks of their life, chicks are not able to maintain their body temperature and must return to the hen at regular intervals for brooding. That is to say, the hen stands or squats with her wings held slightly away from the body and her feathers ruffled. The chicks go under the mother and so benefit from her body warmth. During the first 10–12 days the chicks are in close contact with the hen and brood regularly (Fig. 7.4), whereas later they feed independently from her but still sleep under her. The mother hen is also thought to be important in establishing rhythms of activity and rest for the chicks. Ducks may adopt other ducklings. Under natural conditions this ‘safety in numbers’ probably increases the survival chances of their own young.

Ontogeny and Development of the Young until Independence

Konrad Lorenz is often portrayed being followed by geese that he reared and that were imprinted on him. Young birds of all domesticated breeds have a strong innate tendency to imprint on objects when they first hatch (for a review see Rogers, 1995). This tendency is stronger if the object is approximately the same size as an adult hen, moves and emits sounds. Under natural conditions this object will be the mother hen and imprinting is clearly an adaptive behaviour with strong survival benefits.

Fig. 7.4. Although chicks are precocial, the mother hen is important in the first few days and weeks to help the chicks find food. She also broods them and so helps maintain their body at an appropriate temperature.

All commercial chicks are hatched in incubators and reared in groups. So, although chicks can, and do, survive without a mother hen, there can be problems. As already mentioned, the mother hen helps the chicks to learn about food, but she also helps them to learn to go up to roost on branches at night. Chicks reared without perches for the first 4 weeks of life have greater difficulty learning this behaviour later, and this can affect other aspects of their behaviour in commercial housing systems, such as finding the nest boxes that are often raised up off the ground (Appleby, 1984). Dustbathing and the recognition of dust as a suitable dustbathing material also have to be learnt in the first days of life.

Some simple preening movements are seen on the first day, with others occurring in the following days. The wing tossing in dustbathing is seen on the fourth day. Frolicking, and sparring behaviour is found in young chicks, although aggressive pecking does not become a part of a fight until 3 weeks of age, and aggression is rare until birds are 6–8 weeks old. It starts slightly earlier in males than in females, but in small groups a dominance hierarchy is usually established by 9–10 weeks of age. Those individuals which mature first often tend to become higher ranking than those who develop later, since comb size, which increases rapidly at sexual maturity, is an important predictor of social status. Chickens are not sexually mature until they are 16–18 weeks of age, and they start to lay eggs at 18–20 weeks. Turkeys start to lay eggs slightly later, at 28–30 weeks of age.

Applied Problems

Poultry are the most common of all agricultural species, in addition to being probably the species kept under most intensive conditions. Indeed, it is just this intensive aspect of poultry keeping that has been criticized by consumers. Battery cages, in particular, have been criticized for many decades.

Work investigating several aspect of cage design has resulted in improvements, so that fewer birds become trapped, and horizontal bars in the doors and solid sides to the cages are now used to reduce abrasive feather damage. The main criticism, however, has arisen from more fundamental behavioural research showing the importance to birds of being able to lay their egg in a nest box and to dustbathe in a litter material. Such resources are not included in a standard battery cage, although there is a trend to include them, together with a perch, in so-called furnished cages.

Another development in response to the criticism of the limited space available in battery cages has been the increased interest in loose housing systems for laying hens. Here large numbers of birds are housed in a single room that contains litter, nest boxes and usually perches, although some systems have a raised slatted area. In free-range egg production the bird also has access to an outdoor area. Many of these systems were traditionally used for poultry production, but the modern systems are usually more automated with larger group sizes and higher stocking densities.

The most severe welfare problems in any housing system are those that result in damage, injury or even death of another bird. Cannibalism, when one bird pecks at the skin or underlying tissue of another bird, is one such problem that seems to be increasing in commercial strains of egg-laying birds. The pecking can occur anywhere on the body, but is most serious when it is directed at the cloaca or vent of the bird. It is not unusual to find chickens that have been pecked to death, and since there are often no intestines remaining in the body cavity of these birds, the term ‘peck out’ is often used. In turkeys, cannibalism is mostly seen on the back and wings. Cannibalism rarely occurs in wild fowl and it is not known why it occurs as frequently as it does under commercial conditions in layer flocks. In nature it occurs under adverse conditions often associated with population explosions, food shortages or stress. Feather pecking is when the feathers of one bird are pulled out by another bird (Fig. 7.5). In parrots the bird often pulls out its own feathers. Cannibalism and feather pecking, although both involving pecking, are very different from aggression and aggressive pecking.

Feather pecking is an abnormal behaviour that has not been reported in wild populations of birds. Most researchers agree that it occurs when birds direct pecks to the feathers which, under normal conditions, would have been directed at the ground. It is most likely associated with feeding or foraging motivation, although there are many factors that influence it, including genetic predisposition, light intensity and food composition. Having feathers pulled out is painful and as a consequence of its poorer plumage, the bird has to eat more food in order to maintain its body temperature. The feather-pecked bird is also at greater risk of further injury once the protective covering of feathers has been removed.

Extreme aggression can be a problem under commercial conditions. It usually occurs between males and is a particular problem in quail and turkeys, often leading to wounds on the head and neck. This makes it difficult to keep these species commercially unless under very low light intensities, which in itself can be a problem. Extreme aggression can also be directed towards a few extremely low-ranking individuals within a flock, with the result that they have difficulty in feeding undisturbed and lose body condition.

Stereotyped behaviours are usually related to pacing or pecking. Stereotyped pacing occurs when an animal is frustrated, and is most often seen in association with nesting when access to a nest box is blocked or there is no nest box present. Pecking stereotypies are commonly seen in broiler breeders which, as explained previously, are kept on a severe feed restriction to reduce health and fertility problems associated with high body weight. These birds often develop stereotyped pecking at small spots or flecks on the walls of the pens or polydispsia ('overdrinking').

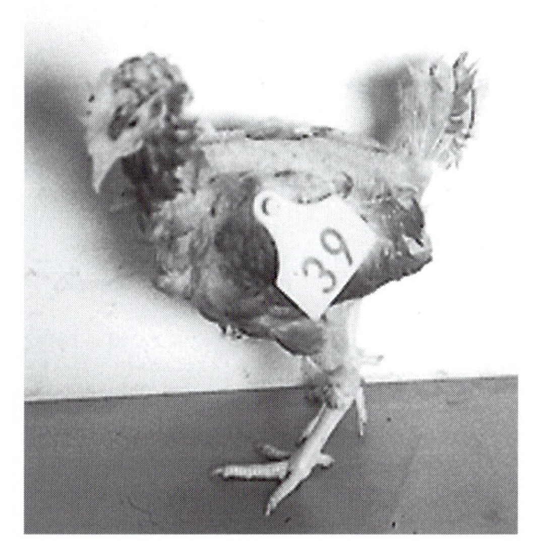

Fig. 7.5. An extreme case of a feather-pecked hen, showing the pattern of feather loss that is typical of feather-pecking behaviour, i.e. naked areas on the back and belly.

A problem that is typical for meat birds, especially broilers and turkeys, is that they have difficulty walking (Weeks *et al.*, 2000). This is a consequence of the intense selection for growth combined with selection for feed efficiency, which has resulted in these birds being less active. Broiler chickens, for example, are much less likely to show perching behaviour, and domesticated turkeys have a reduced tendency to show ground pecking and scratching, and spend a large proportion of the day sitting and resting. The opposite problem to that of inactivity is that of nervousness. This varies considerably between breeds of bird and housing systems, but in large flocks of laying hens in loose housing systems, panic can result in birds rushing to one end of the house, where many are suffocated. Suffocation can also occur in birds that gather by the entrance to the house where the food or the human caretaker usually enters.

Finally, problems in commercial poultry production can be related to birds not using the resources in the intended way, although this is more of a problem for the caretaker than for the birds. For example, hens may not lay their eggs in the nest boxes. Eggs laid on the floor are time consuming to collect and may have reduced hygienic quality. Alternatively nests may be used for roosting in at night and become soiled with droppings, resulting in dirty eggs even if they are laid in the intended place. These so-called behavioural problems are often a consequence of people designing resources based on our perception of what birds want or what we want in terms of price or ease of management. Applied ethology research has much to contribute in the area of poultry housing and design (Appleby *et al.*, 1992). In conclusion, we should be impressed by the flexibility and adaptability of our domestic birds that has led to them being so widespread in agriculture today.

References

Appleby, M.C. (1984) Factors affecting floor laying in domestic hens: a review. *World's Poultry Science* 31, 71–80.

Appleby, M.C., Hughes, B.O. and Elson, A.E. (1992) *Poultry Production Systems: Behaviour, Management and Welfare*. CAB International, Wallingford, UK.

Dawkins, M.S. (1989) Timebudgets in red jungle fowl as a basis for the assessment of welfare in domestic fowl. *Applied Animal Behaviour Science* 24, 77–80.

Duncan, I.J.H., Savory, C.J. and Wood-Gush, D.G.M. (1978) Observations on the reproductive behaviour of domestic fowl in the wild. *Applied Animal Ethology* 4, 29–42.

Leonard, M.L. and Horn, A.G. (1995) Crowing and relation to status in roosters. *Animal Behaviour* 49, 1283–1290.

Mench, J. and Keeling, L.J. (2001) The social behaviour of domestic birds. In: Keeling, L.J. and Gonyou, H. (eds) *Social Behaviour in Farm Animals*. CAB International, Wallingford, UK, pp. 177–209.

Rogers, L. (1995) *The Development of the Brain and Behaviour in the Chicken*. CAB International, Wallingford, UK.

Rushen, J. (1982) The peck orders of chickens: how do they develop and why are they linear? *Animal Behaviour* 30, 1129–1137.

Savory, C.J. and Maros, K. (1993) The influence of degree of food restriction, age and time of day on behaviour of broiler breeder chickens. *Behavioural Processes* 29, 179–190.

Vestergaard, K. (1982) Dust-bathing in the domestic fowl – diurnal rhythm and dust deprivation. *Applied Animal Ethology* 8, 487–495.

Weeks, C.A., Danbury, T.D., Davies, H.C., Hunt, P. and Kestin, S.C. (2000) The behaviour of broiler chickens and its modification by lameness. *Applied Animal Behaviour Science* 67, 111–125.

Wood-Gush, D.G.M. (1989) *The Behaviour of the Domestic Fowl.* Nimrod Press, UK, pp. 18–45.

8 Behaviour of Horses

Sue M. McDonnell

Origin and Domestication of Horses

The ancestral relatives of the horse (*Equus caballus*) evolved as open plains grazers. The early lineage of equids (horses, zebra and asses) has never been clearly established. One of the main questions is whether equids evolved from one common ancestor, or in parallel evolution occurring independently in more than one region of the world.

The first use of the horse by humans was as a hunted wild animal. These large mammals provided a good source of meat and hides. This hunting likely went on for a couple of thousand years before humans started capturing, taming and then breeding horses. It is thought that the domestication process likely started with capture and hand rearing of neonates. Some of the females would be tamed and kept as breeding stock. Once the tamed females matured, there was a clever system for breeding them to wild stallions. The females would be tethered a distance from the village, far enough from people so that wild stallions would approach the mares in oestrus and breed. This was a practical alternative to keeping breeding stallions, which would be both difficult to handle and costly to feed. It is believed that this system of raising and keeping mares and breeding to wild stallions may have carried on in many cultures for centuries before eventually stallions born in captivity were kept for breeding.

Keeping horses provided a handy source of milk and meat. This process of domestication is believed to have occurred about 8000 years ago in Eurasia. Horse bones have been found associated with human settlements in areas that are now Mongolia, Ukraine and Kazakhstan. The earliest evidence of the riding of horses comes from tooth wear suggesting use of mouth bits, dating back about 5000 years.

There remains an interesting controversy about the domestication of horses, with two main competing theories. One theory has been that horses were domesticated in a fairly limited area in Eurasia, so that all domestic horses since then derived from a population of closely related wild horses. It was suspected that domestically bred horses were then spread by humans through Europe, northern Africa and then to the rest of the world. An alternative theory has been that rather than domestic stock being moved from one area to another, the idea and technology of capturing, taming and breeding wild horses spread from one area to another, or perhaps occurred independently. The spread of the process of domestication would mean that the horses we have today derive from a diverse genetic stock rather than from a small population of related horses. Recent methods for evaluating the mitochondrial DNA have been able to begin solving this question. Evolutionary geneticists in Sweden have been evaluating DNA samples from modern horses and DNA from archaeological samples of domesticated horses around the world (Vila *et al.*, 2001). Their most recent evidence suggests that horses were probably not domesticated within a limited area at one time. Rather, the genetic material suggests that domestic horses derive from many different wild horse gene pools, probably spanning a larger geographic area, and also possibly over a longer period of time, than had been believed based on available archeological evidence.

All of the free-ranging or semi-free-ranging horses today are descendants of once-captive or domestic stock. So there are no longer any truly wild horses, as there are wild fowl, sheep or pigs. Many of what we popularly think of as wild horses are domestic horses that after generations of domestic breeding became free ranging. They have fended very well on their own in many instances. Such once-domestic, now free-ranging populations, are known as *feral*.

In addition to feral horses, there are a few remaining herds of a once-wild primitive ancestor of the domestic horse, known as the Przewalski horse or Mongolian horse (*Equus ferus przewalskii*). Several groups of these Przewalski horses are kept around the world, mostly in managed herds confined to large enclosed game preserves (Boyd, 1994). In recent years a small population of Przewalski horse has been successfully re-introduced into wild conditions in their Mongolian area of origin (Fig. 8.1) (Van Dierendonck and Wallis de Vries, 1996).

Social Organization

Horses by nature are social animals that organize into herds. These herds can grow very large, numbering in the hundreds or more, depending on the available resources. Within a herd there are three distinct types of subgroup, called 'bands'. The main family group is known as a harem band. Each harem band consists of one mature breeding male, known as the harem stallion, a few mature breeding females (mares), and their young offspring. The harem stallion stays with his mares and their young offspring all year. The stallion herds and protects the females and young from co-mingling with other bands. The harem stallion drives and directs the movement of the family whenever it is near other bands or when the band is threatened. The harem stallion performs ritualized

Fig. 8.1. Przewalski harem stallion and mares re-introduced from a captive breeding programme to Mongolia in the early 1990s. (Photograph courtesy of Machteld van Dierendonck.)

marking sequences in which he investigates and covers the faeces and urine of his mares by urinating over it. Although the stallion often appears to human intruders as completely 'in charge' of all harem activities, his leadership is limited to defence from intruders. The actual leader of the harem group in terms of everyday quiet maintenance activities is usually a mature mare. So, for example, treks to water, or movement on to a new grazing area, or shifts within the day from grazing to resting, are typically led by a mare.

In contrast to horse stallions that guard a harem of mares and their young, some of the other equids, for example Grevy's zebra (*Equus grevyi*) and asses (*Equus africanus*, *Equus hemionus*), guard a territory. They gain breeding access to females that pass through or linger within the territory.

A relatively rare variation of the single-stallion harem is the two-stallion harem. The second stallion is an unrelated mature male that plays the role of assistant to the main breeding stallion. The assistant is not allowed access to the mature family mares, but may on rare occasion breed with a young mare that is still with the natal band. The main role of the assistant appears to be defence and help with keeping a large harem together.

The young stay with their natal band for 1–3 or more years. The first year is characterized by a large amount of play. This play takes the form of athletic locomotor play, interactive play with the environment, and social play, mostly among the young (Fig. 8.2). The young typically leave their natal band in one of two ways. Some, usually

Fig. 8.2. Common play behaviours of foals and young horses, including play with objects, locomotor play and play fighting. (Drawn by Erlene Michener from photographs by Sue M. McDonnell and Amy Poulin.)

maturing males, are driven off by their sire. This can happen in an abrupt explosive event, or gradually over a period of months. But most appear to leave on their own and gradually. Yearlings and 2-year-olds tend to form affiliations with young from other harem bands. These gangs of young males and females may come and go from their natal band for as long as a year or more before leaving permanently to form new harems.

In addition to the harem bands, a herd of horses includes bands of males that do not have mares. The stallions are called bachelors. They tend to stay together as close to the harem bands as the harem stallions will allow. The young females mature at about 1 year of age and may have their first offspring as early as 2 or 3 years of age. Young males also mature at about 1–2 years of age. The young sometimes have a period of coming and going from the natal band. They may gather with similar-aged young from neighbouring harem bands or hang out near the bachelors during a period of transition from their natal harem band to their eventual adult group.

Within bands, there are often close affiliations. Mares or bachelor stallions often have preferred neighbours within their band. Bachelors often form two- and three-partner alliances to cooperate in controlling limited resources.

In addition to dominance within bands, there appears to be a dominance order among the bands within a herd. Certain bands appear dominant over, or subordinate to, others in terms of access to limited resources such as water, shade and shelter.

Communication

Horses are known for their multi-sensory alertness and instantaneous herd vigilance and reactivity to threat. A variety of postures and expressions appear to be important visual elements of communication. Within and between bands of horses, very subtle changes in ear or tail position appear to convey information. Horses emit a variety of sounds and vocalizations that are likely perceived and serve as communication within the herd. These include specific vocalizations, such as whinnies and neighs, snorts, squeals and grunts. Hoof sounds, from pawing, stamping or contact with the substrate during locomotion also appear to communicate information among horses.

Chemical cues also likely play a large part in communication within and between groups of horses, as well as in perception of threatening predators. Stallions display conspicuous elimination and marking behaviour sequences, in which long periods of time seem devoted to sniffing excretions of herd mates and of faecal pile accumulations of stallions from competing groups within a large herd. Almost whenever herd mates or strangers meet, they approach nose-to-nose with deliberate sniffing of one another's exhaled breath. Many observers have speculated that breath may carry information about relatedness or status.

Among closely bonded band members, particularly mares and their foals, tactile communication no doubt is important. An interesting behaviour that likely includes elements of communication is mutual grooming (Fig. 8.3). Two horses stand facing one another, nibbling insects or tufts of shedding coat from the neck or back of a partner. In addition to obvious grooming needs, this may communicate trust and bonding among participants. Mutual grooming among bachelor stallions or among youngsters often precedes play bouts. So it is believed to be a play-initiation behaviour.

Established order of dominance and submission is an important factor in the social order within and between groups in a herd. An established

Fig. 8.3. Mutual grooming, shown here in week-old foals. (Photograph by Elizabeth Hochsprung Ewaskiewicz.)

order results in very little overt fighting. Most competition is settled with simple threats and retreats. As an open plains species, an important element of avoiding fights is simple retreat from the threat. Very dominant individuals can effectively direct the movement of other animals, or can control a limited resource, with a simple head toss threat or threat gaze that involves a stare with the ears held back toward the neck and the head lowered.

Foraging and Feeding

Horses are grazers that can also browse. They appear to prefer grasses and legumes when available, but will also browse shrubs, woody plants, leaves and roots. For horses living on grasslands, the majority of their time is spent foraging and grazing. Depending upon forage quality, horses under natural conditions typically spend from half to three-quarters or more of their daily time budget grazing. Unlike ruminants, horses may graze continuously for many hours without stopping. When grazing, they move continuously, taking a few bites, taking a few steps forward, taking a few bites. Grazing bouts are usually interspersed with loafing. Depending on the terrain and distance from water, horses may only go to water every couple of days. In wet seasons with lush grasses, horses can obtain sufficient water from the vegetation. In areas where water is freely available, the usual drinking frequency is once or twice a day. Within large herds, the bands usually trek to water in single file, each band with a mare leading and the harem stallion herding at the rear. As the bands approach the watering site, there is typically a fixed hierarchy within each band and between the bands for access to the water.

Biological Rhythms

Under natural conditions, foals are born at the time of year corresponding to best vegetation and survival conditions. In climates with four seasons, most foals are born in late spring. Gestation is just over 11 months. This means that most breeding occurs at more or less the same time of year as foaling. Within about 2 weeks after foaling, mares are again fertile and go through what is known as a post-partum oestrus. Many conceive at this post-partum oestrus. With gestation just over 11 months, this means the mare will foal each year, at about the same time. Foals nurse most intensely from birth to 4–6 months of age. They continue nursing on a less intense schedule until the dam foals again a year later, and some even beyond that. The reproductive life of the healthy well-nourished mare is to foal annually. This means she is pregnant for about 11.5 months, then not pregnant for about 1–2 weeks, and then pregnant again. And a mare that foals every year lactates almost continuously. Horses under natural conditions can live up to 15 or 20 years, and some mares will have foals almost annually from 2–3 years of age.

If a mare does not conceive on the first post-partum oestrus, she will likely have an ovarian cycle of 21 days, with about 2 weeks of dioestrus and 1 week of oestrus until she becomes pregnant. In climates with distinct winter and summer seasons, mares that do not become pregnant in the spring and summer usually become anoestrus during the autumn and winter. This seasonal regulation of ovarian function has been established to be regulated by daylength. As a result, most foals are born in the spring and summer, although exceptions in free-running populations have been noted in which there are examples of foals born in nearly every month.

Courtship and Mating

Under natural conditions, when a mare is in oestrus, the harem stallion stays near, paying increased attention to her frequent urination. Oestrus behaviour includes increased activity, frequent approaches to the stallion, swinging of the hindquarters toward the stallion, lifting the tail and frequent urinations. The mare has a characteristic behaviour involving rhythmic eversion of the vulva exposing the clitoris and lighter internal membranes. This is called 'winking' or 'flashing'. Stallions and mares mate many times a day when the mare is in oestrus. The mare approaches the stallion frequently, signalling her readiness to breed. The female mating stance includes a 'sawhorse squat', with the tail lifted off to the side of the perineum, the head tuned back toward the stallion, and one foreleg flexed. This posture appears to assure the stallion that she will not resist mounting. Most matings occur in less than a minute, and are a relatively quiet event.

Parturition and Parental Behaviour

Gestation in the horse is about 11 months. There is very little change in behaviour with pregnancy or until immediately before parturition. Most parturition occurs in the evening or near dawn. Some especially vigilant stallions appear to sequester their harem off in a sheltered area away from other harems when a mare is foaling. They may continue to keep their distance from other groups for the first days or week after parturition. Parturition is very quick in horses, often only minutes from any sign of discomfort until the foal is born. The mare initially shows signs of abdominal discomfort, getting up and down frequently. Once the fetal fluids erupt, the mare usually remains recumbent for foaling. The family members often gather around the birth event, sometimes vocalizing to the foal during parturition. The mare immediately attends the foal with licking that clears the birth membranes and appears to stimulate the foal. The mare sniffs and may perform the flehmen response to the expelled fetal membranes. Unlike some species, it is rare for horse mares to eat the placenta. Bonding of the foal and dam occurs quite rapidly under natural conditions. Even in harem groups with foals born on the same day, cross-nursing or behaviour suggesting inadequate selective bonding is rarely observed.

Foals are precocious. Those born under natural conditions often stand, nurse and are ready to run with the herd within the first half hour of life. Most of the behavioural elements of the horse repertoire, including grooming, foraging and locomotor sequences, are expressed in either serious or play form within the first day or so of life.

The mare and stallion are protective of the young. For the first week or two the foal's nearest neighbour is almost always the dam, but should any disturbance occur, the harem stallion shows strong protective behaviour towards the young.

Applied Problems

Modern husbandry for domestic horses limits their behaviour. For example, two striking differences are social interaction of horses and feeding. When it comes to social environment, for example, the key to healthy survival for the modern domestic horse is flexibility. Any one horse in its domestic lifetime is likely to experience endless grouping and regrouping, often being exposed to strange horses on a daily or weekly basis. At another time, the horse may be expected to thrive in isolation. Intact stallions, for example, that would be very busy socially under natural conditions, are typically managed on farms so that they have no direct contact with other horses. This is done, of course, to limit aggressive interaction and to control breeding. Horses are also expected in many instances to mix with domestic stock of other species. In general, horses are quite compatible with other domestic species. Horses are well known for their common particular aversion to swine and camelids (camels, llamas, alpacas). Horses not exposed to swine or camelids at a young age may exhibit strong fear reactions when first introduced. Most will eventually acclimate, even as adults.

For practical reasons, very few domestic horses are fed a natural all-forage diet. Instead, it is necessary or more practical to feed horses concentrated grains and other supplements. For many horses this changes feeding duration from many hours spread throughout the day to one or two brief daily meals with long periods of time with little to do. Horse behaviourists cite this change in feeding as the primary cause of behaviour problems of domestic horses, interpreted as arising out of inactivity and 'boredom'.

Common welfare questions for domestic horses include several concerns about humane care and management. Many of these questions concern whether or not certain methods of training and performance are humane. Should punishment and severe training equipment be used, or should behaviour be shaped using only positive and negative reinforcement? Should pain medications and performance-enhancing medications be administered? Similarly, should potentially painful procedures be done to horses for the sake of enhancing appearance or performance? As in dogs and cats, there is also concern about breeding for extreme physical phenotype for show or performance, at the same time predisposing the animal to health problems. Transportation is another area of welfare concern for horses, as it is for all domestic species. Most modern societies are considering limits on duration and conditions of transport. Similarly, there is concern about the duration and degree of work for draught horses. How much social contact and exercise is too little or too much? Related to exercise, there are also growing concerns about housing of horses. How much pasture turnout with natural light, ample space to run, and fresh air is required for the welfare of the animal?

Welfare concerns extend to free-running horse populations as well. In feral horses of North America there are few natural predators. As the horse population increases, available forage is overgrazed and water supplies dwindle. Starving animals encroach on to ranch lands. Population control methods are always politically difficult. While humane destruction on site is likely the most humane, well-meaning citizens attempt capture and adoption programmes that are seen by some as inhumane social and physical stress. In modern society, the use of horses has shifted from primarily agricultural and professional sport, to pleasure hobby horses, often owned by first-generation horse keepers. This brings with it many behaviour and welfare problems related to lack of knowledge and experience with animal husbandry. Related to this, many horses are kept principally for reasons of emotional attachment. One welfare concern is, like our pet dogs and cats, should these animals be required to survive painful injuries and chronic diseases for the emotional benefit of an owner?

Common Behaviour Problems

By far the most common behaviour problems for domestic horses are those related to domestic husbandry incompatible with the nature of the horse. While many horses adjust quite well to isolation or a variety of social and housing and diet changes, others develop stereotypies, separation anxiety or inter-species aggression problems. In general, the more intensely that horses are confined and fed without work or forage to keep them busy, the greater the likelihood of developing such behaviour problems.

By far the most common and troublesome behaviour problems are stereotypies, or repeated movements or stylized behaviour sequences that seem aimless. Estimates of the prevalence of these behaviours among domestic horses range from about 5% to as high as 25%. Stereotypies are rarely, if ever, seen in equids born and kept in natural social and free-ranging conditions. Captive wild-born equids have a very high incidence.

The most common stereotypies are locomotor, including perimeter walking, weaving, pacing and head tossing. A peculiar stereotypy known as 'cribbing' involves grasping on to a surface with the incisors and gulping in and then expelling air through the mouth (Fig. 8.4). Horses also exhibit a form of self-mutilative stereotypic behaviour, typically involving biting of the flank, hind limb or chest, sometimes with kicking against objects.

Another category of behaviour problems of domestic horses involves reproductive behaviour. The specific complaints range from horses showing sexual arousal when people find it inappropriate, to not showing adequate interest in breeding when taken to stud. Most male horses are

Fig. 8.4. Cribbing in a horse. (Photograph by S.M. McDonnell.)

castrated, unless there is expected breeding value. Not all 'geldings', as they are called, lose their interest in breeding. For mares, their performance behaviour can vary with their ovarian cycle. Top-performing mares may exhibit specific undesirable behaviour or may be generally less attentive when they are in oestrus. Maternal behaviour and mare–foal bonding problems are also much more common with intensive domestic management than with mares foaling under more natural conditions.

References and Further Reading

Boyd, L. and Houpt, K.A. (eds) (1994) *Przewalski's Horse: the History and Biology of an Endangered Species.* State University of New York Press, Albany, New York.

Klingel, H. (1975) Social organization and reproduction in equids. *Journal of Reproduction and Fertility, Supplement* 23, 7–11.

McDonnell, S.M. (1999) *Horse Behavior.* Blood-Horse Publications, Lexington, Kentucky.

Mills, D.S. and Nankervis, K.J. (1999) *Equine Behaviour: Principles and Practice.* Blackwell Science, Oxford.

Van Dierendonck, M. and Wallis de Vries, M.F. (1996) Ungulate reintroductions: experiences with the takhi or Przewalski horse (*Equus ferus przewalskii*) in Mongolia. *Conservation Biology* 10(3), 728–740.

Van Dierendonck, M.C., Bandi, N., Batdorj, D., Dugerlham, S. and Munkhtsog, B. (1996) Behavioural observations of reintroduced takhi or Przewalski horses (*Equus ferus przewalskii*) in Mongolia. *Applied Animal Behaviour Science* 50, 95–114.

Vila, C., Leonard, J.A., Götherström, A., Marklund, S., Sandberg, K., Líden, K., Wayne, R.K. and Ellegren, H. (2001) Widespread origins of domestic horse lineages. *Science* 291, 474–477.

Waring, G.H. (1983) *Horse Behavior.* Noyes Publications, Park Ridge, New Jersey.

9 Behaviour of Cattle

Stephen J.G. Hall

Origins of Cattle

Domestication

The wild ancestors of cattle were local races of aurochs (*Bos primigenius*). These were probably domesticated independently around 9000 years ago in western Asia, Africa, China and India (Clutton-Brock, 1999). In India a local subspecies of aurochs was domesticated to give rise to cattle with a shoulder hump, floppy ears and pronounced dewlap, and a characteristic form of Y-chromosome, together with distinctively shaped vertebrae. These are called zebu cattle or *Bos indicus*, and are now common in many subtropical and tropical environments worldwide. The humpless cattle that originated elsewhere are known as taurine cattle (*Bos taurus*). In spite of their genetic differences, taurine and zebu cattle will interbreed freely. Other wild bovine species have been domesticated, namely the banteng, gaur and yak, and there has been some gene flow between them and *B. indicus* and *B. taurus*. The water buffalo and the Cape buffalo are also bovines, but from different genera.

Samples of DNA from fossil aurochs from British archaeological sites have been compared with those from modern cattle (Troy *et al.*, 2001). The fossil specimens were very different from the modern specimens, confirming that today's European cattle are descended from cattle brought from the Near East by the first farmers, and are not the result of local domestication of the aurochs.

Breed differentiation

Today there are 1041 breeds of cattle in the world, of which 217 are rare (S.J.G. Hall, unpublished). In the Western world the trend has been away from multi-purpose breeds, kept in close association with people, and towards breeds specialized for either meat or milk, and farmed as part of intensive systems. In the developing world, cattle serve many purposes in addition to production of food, being used for work, production of manure and provision of financial security. A world database on livestock breeds generally can be found at www.fao.org/dad-is.

Living and extinct relatives of cattle

Feral populations of cattle are not numerous, unlike feral goats or pigs, and most are managed to some extent, mainly by culling of surplus males. The wild ancestor, the aurochs, died out in Britain in pre-Roman times but survived in a game preserve in Poland until AD 1627 (Clutton-Brock, 1999).

Internal Factors Affecting Behaviour

Differences between individuals

Behaviour is partly genetically, and partly environmentally, determined (see Chapter 2). There are differences between breeds in behaviour (Hohenboken, 1986). These probably exist because each breed has been selected to perform well in a particular husbandry system. For example, Friesian (dairy breed) calves are better adapted to intensive veal calf production systems than Salers (beef breed) calves (Le Neindre, 1993), apparently because Friesians have been selected for ability to cope with early weaning and for reduced maternal behaviour.

Within breeds there is individual variation in behaviour, and again this often has a strong genetic component. This has been particularly studied in dairy cattle, where temperament at milking time is very important (Phillips, 1993; Albright and Arave, 1997).

Sex differences in behaviour

Under modern husbandry, from weaning, cattle are usually kept in groups of the same sex and similar age, for example a herd of dairy cows, or a barn or feedlot full of young beef bulls. The behaviour of herds with a natural sex ratio and age distribution has been little studied. Earlier work (Schloeth, 1958) on semi-feral cattle in the Camargue, France, has been supplemented by studies on a small, inbred feral herd at Chillingham in northern England (Hall, 1989). Chillingham bulls and cows show quite marked differences in how they organize their time. During grazing, bulls graze in shorter bouts than cows during the day, but at night bull grazing bouts are longer. On less-productive swards, bulls have shorter grazing bouts. When chewing the cud, cattle can lie or stand; at Chillingham, bulls are more likely than cows to stand to ruminate. This implies that bulls are more on the alert for social encounters, which can happen at any time in this herd, which breeds all year round.

Effects of castration

Castration makes bulls more docile but causes a check in growth. With economic pressure towards rapid growth, in much of Western agriculture castration is being abandoned. However, it is still very widely used in North America, where it is believed to improve meat quality and to reduce the sexually motivated behaviour that can cause much injury in densely stocked intensive beef operations.

Behaviour rhythms

Cattle do not show pronounced seasonal cycles, being able to breed all year round. In practice, they are usually managed so as to calve when there is abundant forage.

The daily rhythm of activity is characterized by alternating phases of feeding and rumination and, in dairy cattle, the timing of milking is also relevant. In traditional pastoral cattle systems, the animals are penned or tethered at night so they can be protected against predators and thieves, but this prevents night-time grazing, which can be very important for cattle (Phillips, 1993).

A herd of cattle usually begins grazing at sunrise; by late morning, most animals are lying ruminating (see also Chapter 10). Grazing may be resumed at midday by at least some animals. This afternoon grazing is more sporadic and some individuals will be ruminating while others graze. As dusk approaches, most or all cattle will be grazing, but this stops soon after dark. Some grazing is also performed at night (Fraser and Broom, 1997). Meals last on average 110 minutes (Phillips, 1993) and there are typically five such meals in a day.

When not feeding or ruminating, cattle exhibit other behaviours, including licking themselves and other cattle (grooming and allogrooming, respectively). This is important for the removal of ticks and as a component of social behaviour.

Foraging and Feeding

Effects of plant and sward structure

When grazing, cattle tend to take the upper layer of the sward, which contains more leaf material. When the sward is short they compensate by lengthening grazing time and increasing bite rate. Cattle graze by gathering a patch of grass into the mouth with an encircling motion of the tongue; the tongue and the lower incisor teeth pin the grass against the upper palate and a head movement breaks the grass off. They can graze at up to 70 bites per minute (Phillips, 1993; Fraser and Broom, 1997). Like goats and sheep, and the other ruminants, cattle do not have incisors in the upper jaw. Cattle cannot graze closer than 1 cm to the ground, and are relatively unselective. As the sward is eaten down, cattle tend to maintain rate of intake, by accepting herbage of poorer quality, while sheep tend to maintain quality of intake, by reducing bite size and bite length (references in Dumont *et al.*, 1995). In contrast, sheep and goats have a relatively mobile upper lip which helps them to select individual leaves from a grass plant.

Why have the upper incisors been lost in ruminants? Scott and Janis (1993) suggest this happened in parallel with the lengthening of the ruminant jaw, and with the development of the tongue for the prehension of vegetation.

Eating forages and concentrates

The feeding behaviour of cattle given silage, hay, pellets, other conserved feeds or cut browse or herbage, has some similarities with grazing behaviour. There is a meal structure, and with hay and silage cattle show some selectivity. However, because of the crowded conditions normal for housed cattle, there can be accentuated competition between cows for access to feed. When grazing is supplemented by feeds of this kind, cows will often cut down their grazing behaviour and allocate the time thus saved to lying or resting behaviour (Phillips, 1993).

Drinking

Zebu cattle in Africa have been recorded as drinking 104 kg of water (28% of the dehydrated body mass) in 4 minutes, and they can be managed so they only need to drink every 3 or 4 days, even during the dry season (Nicholson, 1985). Mean daily water intake of a cow in Europe during the period May–September is 25.1 kg (Castle and MacDaid, 1975). Major factors influencing intake are the dry matter content of the herbage, rainfall, maximum air temperature and, to a lesser extent, atmospheric relative humidity and the number of hours of sunlight. However, milk yield does not have an important influence.

Rumination

Rumination enables cattle and their relatives to exploit fibrous vegetation, by providing an environment (the rumen) for symbiotic microorganisms. After a meal, material is regurgitated, chewed and swallowed again. Rumination corresponds to about three-quarters of the time spent grazing (Fraser and Broom, 1997). It occupies about 6–7 hours of a cow's day; bouts are about 45 minutes in length and, while chewing the cud, alertness is reduced and cattle will become drowsy. They may lie down, or remain standing, to ruminate.

Social Behaviour

Relationships between animals

Relatively few herds of husbanded cattle include large numbers of adult bulls, so our knowledge of bull social behaviour tends to come from feral herds. Here, bulls have a home range system and a more strictly linear hierarchy than cows (Hall *et al.*, 1988). In all herds of adult cows, dominance structure is relatively stable from year to year and it can be highly complex, with individual cows sometimes consistently dominating cows who appear to be of higher social rank. Particular cows form 'friendships' which last for long periods of time. It is thought that a cow can recognize 50–70 other individuals (Fraser and Broom, 1997). Cattle lick each other, and this allogrooming has been suggested as a comfort mechanism, helping cattle to cope with intensive husbandry systems (see references in Phillips, 1993).

Herd structure and dynamics

Free-ranging cattle tend to move around their habitat as groups of cows and calves. Bulls may join and leave these groups; they often associate in small bull groups, and each such group usually inhabits a particular area (Hall *et al.*, 1988; Hall, 1989). In husbanded, as well as free-ranging cows, there is a social hierarchy which is established and maintained by social interactions. It is most obvious when cattle are closely confined, when individuals will be seen to differ in whether they move out of the way of particular other individuals. The hierarchy is based on the dominance–subordination relationships that exist between each animal, and each other individual in the herd. About 25% of these relationships may change each year, in that a cow that used to be subordinate to another may become dominant to her (Phillips, 1993). This social hierarchy, based on spatial relationships, is usually also seen when animals are feeding in close proximity to each other. However, it does not apply to all aspects of social life. The well-known and predictable order of entry into the milking parlour is not strongly related to social hierarchy; it tends to be the higher-yielding cows that enter first. During grazing, it is not necessarily the most dominant cows that lead the herd (Phillips, 1993).

Spacing between animals is influenced by many factors, including herbage conditions, and cattle will tend to group together when flies are particularly annoying.

Communication

Visual signals

These are most evident with bulls, who paw the ground, rub head and neck in the earth, thrash vegetation with the horns, vocalize, and exhibit a wide range of displays, mainly based on positioning the body in various ways to accentuate the size and strength of the shoulders (Hall, 1989). In the lateral display, bulls stand parallel to each other but facing opposite ways (the 'inverse polar' position) and may turn their heads slightly away from each other, the front of the head being vertical to the ground. These displays may culminate in a fight, which involves stabbing movements of the horns, and head-to-head pushing, especially in the case of hornless cattle. These behaviours may also be shown by cows, but in a much less overt manner. Facial expressions of cattle are very limited and are limited to the 'lip-curl' of flehmen, and to the mouth opening that accompanies vocalization. The ears do not really have an expressive function. Body attitude is important and is defined by the angles that the head, neck and line of the back make with each other. The tail seems not to be important in communication, except that young calves often wag the tail while suckling, and may hold it up while running during play. Also, oestrous females will move the tail around in a restless manner.

Some of the main signals conveyed by body attitude are summarized in Fig. 9.1.

Olfactory signals

Bulls sniff the urine and the perineal region of cows, and can detect a change in the cow's hormonal state up to 4 days before the start of oestrus (Phillips, 1993). After sniffing, the bull may or may not perform flehmen – the characteristic 'lip-curl' expression, with the head held up (Fig. 9.2), which aids the perception of pheromones (Hall, 1989; and Chapter 10). Cows can identify their calves by smell, although recognition by sight and by sound becomes more important as the calf grows up; adult cattle will still sniff each other during social behaviour (Phillips, 1993).

Vocalizations

Vocalizations have been described in husbanded cattle, but the social contexts in which they are used have not been studied critically, except for the cow–calf contact calls which are so evident when a calf is removed from its mother.

The cow's vocalization appears to be essentially the same as the calf's, an undifferentiated 'moo' sound, but with greater volume and lower pitch. Vocalizations of adult bulls have not been much described; modern breeds have, perhaps, too much muscling around the neck region

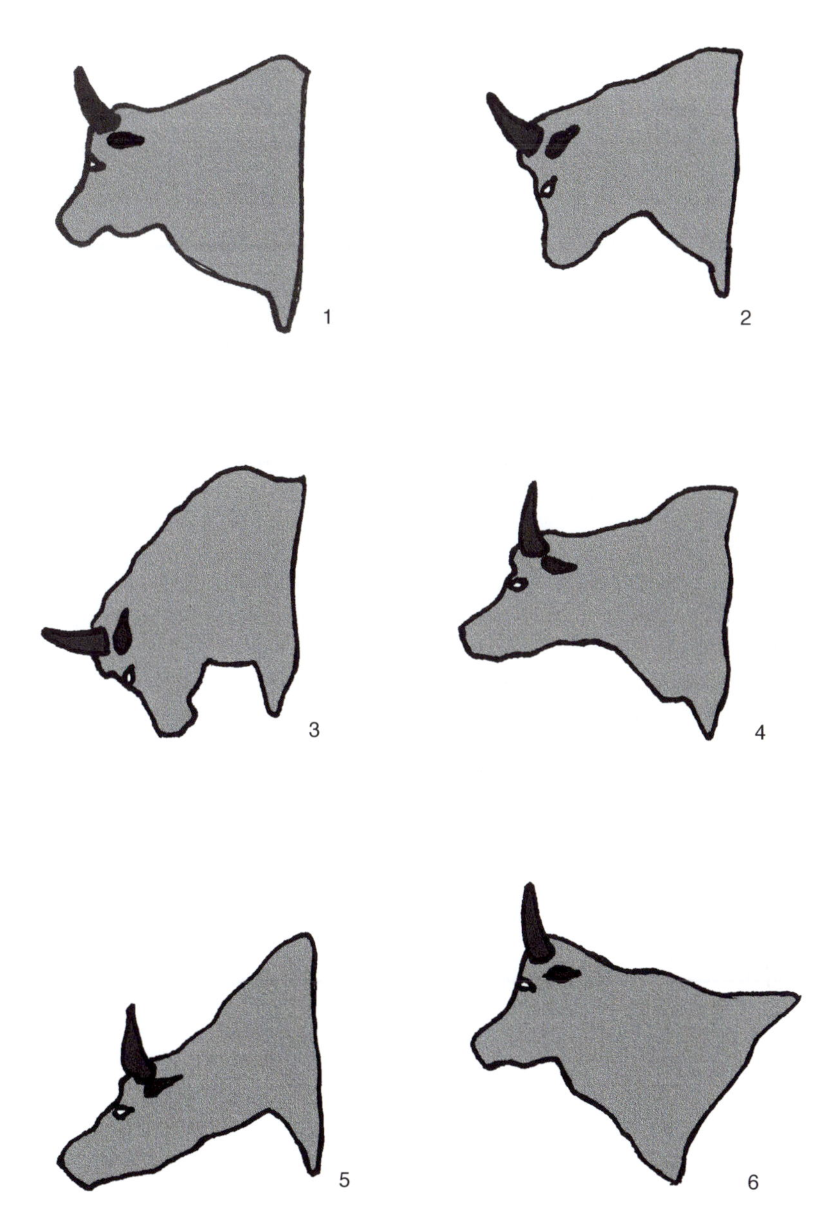

Fig. 9.1. Positions of head and neck in adult cattle and the signals they convey. 1, Normal (neutral) position; 2, lateral display, low intensity; 3, lateral display, high intensity; 4, approach, relatively confident, to another animal; 5, approach, submissive, to another animal; 6, alert position; line of back concave, often followed by running away. (Redrawn from Schloeth, 1958.)

to permit the vocal apparatus to resonate and function fully. Camargue and Chillingham bulls (Schloeth, 1958; Hall *et al.*, 1988) have at least two vocalizations, a 'low' (or 'moo') and also a 'call', 'hoot' or 'roar' which is of higher pitch, and consists of a series of repeated, brief calls.

Fig. 9.2. Bull performing flehmen on the urine of a year-old heifer. These cattle are of the Lincoln Red breed, a traditional English beef breed.

Mating Behaviour

Oestrus

As the time of ovulation approaches, the cow behaves in an increasingly agitated manner, which presumably helps to attract bulls. Cow–cow mounting is very obvious in dairy herds; when a bull is present, this behaviour appears to be suppressed, partly because the bull will guard the cow from approaches by all other cattle. Bulls will guard a cow as she approaches oestrus. This guarding is an active process – as well as chasing off other bulls, he will prevent her from rejoining the herd, so she remains behind with him as the herd moves on. Periods of active guarding are interspersed with periods when the cow stands still, and then the bull usually stands in the inverse polar position, parallel and close to the cow, but with his head by her rump.

When she becomes receptive, which may be 24 hours after the start of oestrus, the bull may serve her about five times; when there is no bull, as in herds which rely on oestrus being detected by the stockman and on artificial insemination, female–female mounting is very evident. About 90% of all mounted cows are in oestrus, but only 70% of all mounting cows (Phillips, 1993).

Maternal Behaviour

Giving birth

The cow will leave the herd before giving birth, although if it is her first calf, she may not do so and the calf may be born in the middle of the herd. The calf may be born while the cow is standing or when she is lying; fetal membranes are eaten and the calf is usually licked intensely immediately after birth, although some cows rest before licking starts. Some studies have found a tendency towards night-time calving; others have not, while some have shown how cows may avoid calving at milking time (Fraser and Broom, 1997). The calf usually succeeds in standing within the first 30–60 minutes and teat-seeking behaviour follows; this is often less successful in calves born to cows with fat teats or very large udders. The cow will return to feed her calf perhaps twice a day, and after a few days instead of lying down after being fed the calf will follow its mother back to the herd (Phillips, 1993). In this 'lying-out' behaviour calves resemble kids and are very different from lambs and foals, which follow their mothers back to the herd soon after birth.

The milk-ejection reflex

Suckling stimulates nerve endings in the udder, and the nerve impulses travel through the spinal cord to the posterior pituitary gland, which releases the hormone oxytocin. When this reaches the udder through the bloodstream, the cells around the milk-containing vessels (the alveoli) contract and milk is ejected. This reflex can be conditioned; many cows will only show it if their calf is present, while others may begin to let down their milk when they notice the beginning of the milking routine. In many breeds, especially of zebu cattle, the presence of the calf is essential for milk to be let down (Phillips, 1993).

Nursing, suckling and weaning

During suckling the calf is usually in the inverse polar position, gaining access to the udder from in front of the hind leg (Fig. 9.3). The calf moves rapidly from teat to teat. Suckling bouts are 10–15 minutes long and the number per day is initially 5–8 (newborn) declining to 3–5 later (Phillips, 1993). Bouts may be initiated by the cow or the calf; one may call, the other responds, and the two will rapidly find each other and begin suckling. There may be an imitative component; the sight of a calf being suckled often appears to encourage other cows to seek their calves and suckle them too.

Fig. 9.3. Chillingham cow suckling her calf.

Calves are suckled about 5–8 times each day when very young; this declines as the calf grows. Whether male calves receive more milk than females is not known. There is a clear effect of calf body weight on milk production in natural suckling, and as male calves are heavier than female calves of the same age (Somerville *et al.*, 1983), a greater maternal investment in male calves seems likely. Natural weaning is seldom seen; Reinhardt and Reinhardt (1981) found female calves were weaned by their mothers at 8.8 months of age, and males at 11.3 months.

Behavioural Development of Young

Learning and the development of behaviour

Young calves are usually very playful (Fraser and Broom, 1997). The most prominent behaviours are running and prancing about, with the tail usually held high. Playful fighting, head-pushing and mounting are commonly observed in calves of all ages, although their incidence and the ages of onset of the different behaviours probably vary widely between herds. Calves learn to graze effectively, as shown by the increase in bite rate as they grow up. Cows can learn specific tasks quite quickly. This is partly due to their lack of timidity and willingness to explore and investigate unfamiliar areas and objects (Phillips, 1993).

Development of social and sexual behaviour

In modern dairy farming the calf is usually removed from its mother 24 hours after birth, to be reared on artificial milk replacer for 5–7 weeks. The cow–calf bond is not completely formed within the first 24 hours of contact, so it is possible that this early separation is less stressful than a separation after a longer contact (Le Neindre, 1993; Phillips, 1993). Calves that are subsequently kept in individual pens (for disease control) do not have the opportunity to play, and group-rearing of these young calves leads to more complete behavioural development and is, for this reason, much better for welfare. Play behaviour and the opportunity to associate with other young calves are of clear importance to behavioural development.

Under husbandry conditions, social behaviour can be influenced by imprinting on to stockmen (Phillips, 1993). This is especially obvious in pastoral societies, where livestock husbandry is highly developed, for example Lott and Hart (1979) describe how, among the Fulani people of West Africa, stockmanship is highly regarded, and the stockmen establish and continually reassert their dominance over each individual in the herd.

Development of feeding behaviour

Cattle develop their grazing skills as they grow (Phillips, 1993). Calves of 8 weeks take 14 bites per minute, but at 18 weeks the rate is 50 bites per minute, nearly as high as that of adult cattle. Short-term and long-term memory play their parts too; cattle learn to avoid toxic plants and they have the capacity to select a diet which is appropriate to their current nutrient requirements, by means that are not yet understood. They appear to monitor their rate of food acquisition and use a change in this rate as the cue to move to another feeding site.

Welfare Issues

Concepts of animal welfare and its assessment are discussed in Chapter 6; for practical purposes, whether welfare is good or not in a given situation is often assessed in terms of whether the animal is able to express the five freedoms (Chapter 6). Very many of the practices of modern cattle husbandry challenge at least one of these freedoms. The high-yielding dairy cow is under nutritional stress; her environment, especially in cubicle housing, is often not appropriate because of increased likelihood of injury and lameness, while normal behaviour cannot be expressed unless she has adequate access to grazing and is in a stable social grouping. Full expression of cow–calf interaction is not usually possible. In contrast, extensively farmed beef cattle probably have among the best welfare of any farm animal, by the criterion of accordance with the five freedoms.

Calves which are separated from their mothers and are reared for veal or beef exhibit many abnormal behaviours. When young they often show inappropriate non-nutritive sucking behaviours, and when rather older, excessive mounting behaviour, which often causes injury and retarded growth (Phillips, 1993; Fraser and Broom, 1997).

The welfare of dairy cattle has been studied extensively (Phillips, 1993). Linkages between patterns of behaviour and incidence of disease have been explored – socially subordinate cows are more likely to become lame, apparently because they are forced to spend time in wet parts of the shed. Dairy cattle are usually housed for at least part of the year and some for all of the time. Stocking density is thus artificially high and the opportunities to perform the full range of behaviours are limited. Experimental studies have revealed the motivation of cattle to perform specific behaviours. It is now known, for example, that cows have a strong motivation to spend part of each day lying down, and that they will sacrifice feeding time in order to do so.

The cow–calf bond does not develop immediately on birth of the calf, so early separation of the pair may be better for welfare than delayed (e.g. 24 hours post-partum) separation. Among the most widely publicized animal welfare issues is the rearing of calves for veal in small crates, which has been banned in several countries. Welfare concerns regarding the cattle industries are likely to increase as public awareness of the negative aspects of livestock intensification grows, together with the commercial pressure for such intensification. Even apparently benign developments such as robotic milking (which allows the cows to choose when to be milked, reducing stress on the udder) could lead to less attention being paid to the animals (Rollin, 1995), and the same may also apply to extensified beef production.

References and Further Reading

Albright, J.L. and Arave, C.W. (1997) *The Behaviour of Cattle.* CAB International, Wallingford, UK.

Castle, M.E. and MacDaid, E. (1975) The intake of drinking water by dairy cows at grass. *Journal of the British Grassland Society* 30, 7–8.

Clutton-Brock, J. (1999) *A Natural History of Domesticated Mammals*, 2nd edn. Cambridge University Press, Cambridge.

Dumont, B., Petit, M. and d'Hour, P. (1995) Choice of sheep and cattle between vegetative and reproductive cocksfoot patches. *Applied Animal Behaviour Science* 43, 1–15.

Fraser, A.F. and Broom, D.M. (1997) *Farm Animal Behaviour and Welfare*, 3rd edn. CAB International, Wallingford, UK.

Hall, S.J.G. (1989) Chillingham cattle: social and maintenance behaviour in an ungulate which breeds all year round. *Animal Behaviour* 38, 215–225.

Hall, S.J.G., Vince, M.A., Walser, E.S. and Garson, P.J. (1988) Vocalisations of the Chillingham cattle. *Behaviour* 104, 78–104.

Hohenboken, W.D. (1986) Inheritance of behavioural characteristics in livestock. A review. *Animal Breeding Abstracts* 54, 623–639.

Le Neindre, P. (1993) Evaluating housing systems for veal calves. *Journal of Animal Science* 71, 1345–1354.

Lott, D.F. and Hart, B.L. (1979) Applied ethology in a nomadic cattle culture. *Applied Animal Ethology* 5, 309–319.

Nicholson, M.J. (1985) The water requirements of livestock in Africa. *Outlook on Agriculture* 14, 156–164.

Phillips, C.J.C. (1993) *Cattle Behaviour.* Farming Press, Ipswich.

Reinhardt, V. and Reinhardt, A. (1981) Natural sucking performance and age of weaning in zebu cattle. *Journal of Agricultural Science, Cambridge* 96, 309–312.

Rollin, B.E. (1995) *Farm Animal Welfare. Social, Bioethical, and Research Issues.* Iowa State University Press, Ames, Iowa.

Schloeth, R. (1958) Cycle annuel et comportement social du taureau de Camargue. *Mammalia (Paris)* 22, 121–139.

Scott, K.M. and Janis, C.M. (1993) Relationships of the Ruminantia (Artiodactyla) and an analysis of the characters used in ruminant taxonomy. In: Szalay, F.S., Novacek, M.J. and McKenna, M.C. (eds) *Mammal Phylogeny.* Springer-Verlag, New York, pp. 282–302.

Somerville, S.H., Lowman, B.G., Edwards, R.A. and Jolly, G. (1983) A study of the relationship between plane of nutrition during lactation and certain production characteristics in autumn-calving suckler cows. *Animal Production* 37, 353–363.

Troy, C.S., MacHugh, D.E., Bailey, J.F., Magee, D.A., Loftus, R.T., Cunningham, P., Chamberlain, A.T., Sykes, B.C. and Bradley, D.G. (2001) Genetic evidence for Near-Eastern origins of European cattle. *Nature* 410, 1088–1091.

10 Behaviour of Sheep and Goats

S. Mark Rutter

The Origins of Sheep and Goats

Sheep and goats were two of the first species to be domesticated by humans. The domestic breeds of the two species are believed to have originated from wild animals living in dry and mountainous regions in south-west and central Asia between 8000 and 10,000 years ago. The domestic goat (*Capra hircus*) is mainly (if not entirely) derived from the bezoar goat, (*Capra aegagrus*). It is now believed that all the domesticated sheep breeds (*Ovis aries*) were derived from the Asiatic mouflon, *Ovis orientalis* (Clutton-Brock, 1999). The European mouflon (until recently called *Ovis musimon*, but now generally included in *Ovis orientalis*) is now believed to be a relic of the first domestic sheep rather than a relic of a separate wild species. Given that sheep and goats were (and still are) often run together, their spread into Europe, Africa, the Americas and Oceania occurred largely in parallel. Sheep were (and still are) valued mainly for their meat and wool, whereas goats are valued mainly for their meat and milk.

Domestication history

The fact that the goat and sheep feature widely in Greek mythology is evidence of their importance at this time. A goat named Amalthea raised old Zeus, and he rewarded her by placing her in the sky as the star Capella, or 'Little Goat' (Mowlem, 1996). In 400 BC the Greeks formalized the Zodiac, which included the goat–fish Capricorn and the ram Aries. By the first century AD, the goat was beginning to be discriminated against, with Christ's followers being described as 'good sheep', in contrast to the others, the 'goats'. The changes in the relative agricultural importance of the two species over the past two millennia have mirrored this early Christian discrimination. The fact that sheep are tolerant of cold, wet climates helped their spread from the Middle East. The Romans introduced a white-woolled sheep into western Europe to meet the need for woollen clothing in their northern provinces. The predominance of the sheep in the West continued throughout the Middle Ages, and now, in the developed countries, sheep production (and the number of animals reared) is very much greater than goat production. Indeed in New Zealand, sheep production (primarily wool) is of significant economic importance, with sheep outnumbering the human population by a factor of 12. About three-quarters of the goats in the world are now found in the developing countries. However, in many of these countries, the goat is still the most important source of protein, and many communities are dependent on their flock of goats.

Relative economic importance

The fact that there are fewer goats in the world than sheep, and that the majority of goats are kept in the developing countries, has led to the majority of research and development in the two species being focused on sheep. This imbalance is reflected in the scientific literature covering the two species. However, the virtues of the goat, especially its tolerance of hot, humid environments, are now leading to efforts to increase the contribution of the species to modern agriculture.

Foraging and Feeding

Digestive physiology

Sheep and goats are both ruminants (as are cattle, see Chapter 9). That is, their stomachs are divided into four separate compartments, the largest of which is called the rumen. This contains a variety of microorganisms that are capable of digesting cellulose. This means that sheep and goats (like cattle) can eat and digest grass and other plant material. Rumination (commonly known as 'chewing the cud' or 'cudding') forms an important part of the digestive process of ruminants. This involves partially digested material from the rumen travelling back up the oesophagus to the mouth. Here it is chewed again, typically for about 1 minute, before being swallowed. This additional chewing helps with the microbial and protozoal breakdown of the plant cellulose. Both animal species typically spend up to one-third of the day ruminating. Although sheep and goats share the same digestive physiology, the two species show important differences in their foraging behaviour.

Goats as browsers

Goats are predominantly 'browsers'. That is, they will eat the leaves and shoots of trees and bushes. This does not mean to say that this is all that they eat, just that this can form part of their foraging behaviour when the animals are given this option. Research has shown that goats prefer to browse even when grass is abundant, and that, given a choice, browse forms between 50 and 80% of the diets of goats. However, goats can happily be kept where their only foraging option is to graze grass. There are two factors that determine the goat's ability to browse. First, the upper lip of the goat is more mobile than that of the sheep, allowing the goat to be more selective, and to be able to 'pick' preferred plant parts with relative ease. Secondly, the efficiency with which goats can digest coarse roughage, such as the leaves of many trees and shrubs, is greater than that in sheep. The fact that goats select their diets from a greater range of forage sources encourages them to wander over longer distances in the search for food. This leads to gaps in their eating activity as they move from one tree or shrub to another. This extra searching time means that goats typically spend up to 11 hours a day browsing, whereas sheep typically spend 8 hours a day grazing. Goats have naturally inquisitive feeding habits, and will often try to eat inappropriate substances. This is demonstrated by the tales of goats eating articles of clothing that have been inadvertently left in their reach!

Sheep as grazers

In contrast, sheep are predominantly 'grazers' (Lynch *et al.*, 1992). That is, they eat grass (or other herbage). Sheep, in contrast to goats, do not relish eating the leaves of trees or bushes, and will eat grass when given a choice. However, this does not mean that sheep do not have dietary preferences. Research has shown that sheep prefer to eat clover when given a choice between grass and clover (see Fig. 10.1 for an explanation of how preference is measured). However, this is not a total preference, and, given a free choice, sheep will eat approximately 70% of their diet as clover (Parsons *et al.*, 1994). Given that sheep can eat clover more quickly than grass, the fact that they eat any grass at all is unusual. That is, if they were 'rate maximizers', they should only eat the food they can eat more quickly, i.e. the clover.

Diet preference theories

There are mainly two possible explanations for why sheep eat some grass (Rutter *et al.*, 2000). The first is based on the fact that there is a clear diurnal pattern to preference. Sheep show a strong preference for clover in the morning, with the preference for grass increasing towards the final meal of the day, just before sunset. Sheep, as with other ruminants, avoid grazing at night, and this is believed to be an innate antipredator response. The passage rate (i.e. the speed with which food passes through the digestive system) for clover is quicker than for grass. If the sheep ate just clover, the quick throughput increases the likelihood that they will have to have a meal during the hours of darkness. In contrast, if they eat grass in the evening, the slower throughput

Fig. 10.1. Diet preference for grass and clover is measured by observing (from the tower) the proportion of time the animals spend, when given a free choice, grazing from a pure sward of clover (in the foreground) alongside a pure grass sward (in the background).

decreases the likelihood that they will require a meal at night. The second possible explanation of why sheep eat some grass is associated with their ability to cope with dietary changes. Grass contains a higher proportion of fibre (in the form of the structural carbohydrates cellulose and hemicellulose) than clover. If the sheep ate only clover, the microbial and protozoan population living in its rumen would undoubtedly change, such that those organisms capable of digesting cellulose would decrease. This would reduce the efficiency with which the animal could digest grass, and could place the sheep at a competitive disadvantage to other sheep that had maintained grass in their diets, should clover no longer become available. In other words, by maintaining some grass in its diet, the sheep can maintain its ability to cope with a change in its diet should circumstances dictate.

It is not known at this stage which of these two hypotheses is the correct explanation. It is quite possible that both processes influence the sheep's foraging behaviour. These theories demonstrate that ethologists believe that much of the behaviour of modern domestic animals is governed by innate patterns of behaviour that were developed in the evolutionary history of the species. Even though humans have protected, managed and selected these animals for centuries, these innate behaviour patterns from their wild ancestors are still present in modern sheep and goats, just as they are in the other domestic animals described in this book.

Impact of foraging on the environment

The differences in the foraging behaviour of the two species allow them to utilize different types of vegetation. This complementarity is one of the reasons why sheep and goats are often kept together. The fact that goats can, and will, eat a variety of plants that other herbivores will not touch (e.g. thistles and nettles) means that they are often used to control these plants (see Fig. 10.2). However, the fact that goats can, and will, eat a wide variety of vegetation has led to the reputation of the species for deforestation and desertification around the world. This followed the problems resulting from the establishment of uncontrolled, feral goat populations in many countries and islands around the world. Some of these populations were the result of deliberate introductions to eat brush and undergrowth, while others were escapees from domestic herds. However, the goat's reputation as a 'deforester' is undeserved, and the problems can usually be attributed to the mismanagement of humans. The fact that the goat, which can find sustenance where other species cannot, is often the last to remain in such situations appears to have led to it taking the blame.

Social behaviour

Both sheep and goats live in small groups (or flocks). This provides advantages to the individuals within the group, including protection from predators, better conditions for the survival of the young and access to mates. Note that although the farmer usually keeps all of his sheep together in one flock, the sheep within this flock may form themselves into separate, discrete subgroups. Each subgroup has its own home range within the overall flock home range. This is why flocks from different farms can be kept on the same area without the need for a fence to separate them, as the animals naturally keep in separate groups (or 'hefts'). This is passed from generation to generation, with the lambs of a ewe occupying the same home range as their mother.

Social Interactions

Both goats and sheep show a strong desire to remain with their group-mates, and become very vocal when separated from the flock. Both species use threat behaviour to help minimize fighting between individuals. This helps to minimize the risk of injury associated with fighting. Threat behaviour is similar in both sheep and goats, and consists of a lowered head and a stretched neck. If the threat fails to deter a potential rival, the males of both species engage in rearing and butting to establish their dominance within the group. Sheep approach each other, and butt head on. In contrast, goats stand about 1–2 metres apart, then rear up with their body at right angles to their opponent. They

Fig. 10.2. Goats can be used to control unwanted vegetation. This rush infestation (top) was eliminated (bottom) after 3 years of goat grazing.

then pivot, lunge forward and come together with a loud crack. This difference in the agonistic behaviour between the males of the two species means that sheep and goats can be kept together with little conflict between them. Although the majority of the animals in a group will use aggression at some time, up to 25% of the individual relationships between goats in a group can be peaceful, i.e. do not involve any aggression. The fact that individual sheep and goats can avoid more dominant group members is strong evidence that they can recognize other individuals within their group. Although visual cues are no doubt important in individual recognition, the fact that sheep and goats spend a lot of time in mutual sniffing, especially with a strange animal that is introduced into the group, implies that olfaction must also be important in this respect.

Social Spacing

Sheep within groups maintain a certain distance from their nearest neighbour when grazing. This nearest neighbour distance tends to be a characteristic of the breed, with hill sheep breeds usually found further apart than lowland breeds. Nearest neighbour distance also decreases as vegetation quality and homogeneity increase. In Scottish hill sheep, each of the clearly defined cohesive groups will monopolize the use of an area of hill and will avoid close social contact with sheep from other groups. The area utilized by the group is at a maximum in the summer and a minimum in the winter. The expansion of the area used by the group in the summer was shown to be as a result of the behaviour of only a few flock members. These more independent members of the flock appear to be less concerned with maintaining close contact with other sheep. These sheep are also responsible for the direction of movement of the sheep around the paddock. The remaining sheep follow in several files that join together as they move towards the same point.

In cold, wet weather, sheep will huddle together to afford mutual shelter and to conserve body heat. Goats living in hot, dry and treeless areas will also congregate and huddle during the midday heat. Although this latter behaviour seems to go against common sense, it occurs when the heat being taken in by the goat's body exceeds its ability to dissipate heat. By huddling together, the goats can reduce the intake of direct and reflected solar energy, and so are better able to maintain their body temperature than if they were stood alone in the sun.

Mating Behaviour

The mating behaviour of sheep and goats is very similar (Hafez and Hafez, 2000). As with other mammals, the courtship displays of sheep and goats are comparatively simple, and do not involve the complex, ritualized displays found in birds, fish or arthropods. As with other domestic animals, contact between sexually receptive females and males of the two species is usually managed by the farmer. Both sheep and goats are seasonally polyoestrus, with recurring oestrus periods in the autumn breeding season. The libido of both sexes of both species is generally low outside the breeding season. The males become more aggressive during the breeding season. The average duration of the oestrus cycle is 16.7 days for sheep and 20.6 days for goats. During oestrus, which generally lasts between 18 and 24 hours, female sheep and goats show increased motor activity and appear restless. Receptive females show an increase in the frequency of non-specific bleats. Nanny goats tend to mount and be mounted by other females, although this is exceptional in the ewe. The males of both species use olfactory and gustatory stimuli to detect females that are in oestrus. The male smells the female's urine and then stands rigid, with his head raised and lips curled. This is known as 'flehmen' behaviour, and typically lasts between 10 and 30 seconds in male goats and sheep. The female will also sniff the male's body and genitals at this stage of mating. This head to genital approach of both the male and female leads them to circle each other. The male makes courtship grunts and licks the female's genitalia during sexual approach. Male goats will frequently urinate on their forelegs during sexual excitement. The male will nudge the female, and she will turn her head back towards the male. Courtship terminates when the female becomes immobile and adopts a posture that allows the male to mount. If the female is receptive, copulation can occur rapidly, and is very brief. The male's head moves backward rapidly at ejaculation. The male then dismounts.

Post-coital behaviour

After mating, rams will stretch their head and neck, and goats will lick their penis. Most males show no signs of sexual activity for a period after mating (known as the refractory period). In dairy goats, watching another male copulate enhances the sexual performance in the spectator when he is subsequently allowed to copulate. One unique characteristic of the goat is that, under certain circumstances, the sudden introduction of a male in the period between the breeding and non-breeding seasons can synchronize the oestrus of a group of females. Although natural mating between the two species can lead to conception, the embryos usually die at 30–50 days.

Birth and Parental Behaviour

Maternal behaviour around birth

Gestation in both sheep and goats normally lasts for 149 days. Variation among breeds is reported from 143 to 151 days, although individual pregnancies can last from 138 to 159 days. Both ewes and nanny goats will leave the flock to give birth. In the ewe, actual parturition (after dilation of the cervix) takes about 15 minutes per lamb. The umbilical cord is broken by stretching. If not already standing, the mother will stand within 1 minute of giving birth. Whether or not the mother eats the fetal placenta (or afterbirth) depends on the breed. Immediately after birth, the vigorous licking by the mothers of both species has a stimulatory effect on their offspring. Nanny goats are particularly vigorous in grooming and orienting to their first-born. This means that the second-born, which is usually the weaker of the two, has a greater opportunity to suckle. It is during this phase of intense licking, which lasts for about an hour, that the ewe learns to distinguish her lamb from others. Vigorous head butting by the ewe will reject alien lambs. If the lamb is removed from its mother during this critical phase following birth, the strong maternal bond can be broken and the lamb will be rejected if presented to her 6–12 hours later. This process of maternal bonding is much quicker in the goat, only requiring a few brief moments after birth. Grooming by the mother is also important in that it removes much of the 500 g of fluid that is present in the coat of the offspring at birth. This helps reduce heat loss, as young lambs and kids are both particularly susceptible to chilling. Should the lamb die, the mother rapidly loses interest in the cold, immobile body.

Cross-fostering

It is possible to cross-foster lambs and kids. That is, ewes can be persuaded to adopt goat kids, and nanny goats can be persuaded to adopt lambs. However, for the procedure to be successful, it must be carried out as soon as possible after birth, especially for the nanny goat, and certainly during the critical 1-hour period following birth for the ewe. The procedure is similar to that used to get a mother of either species to adopt offspring of their own species (i.e. fostering). The lamb or kid to be introduced should be smeared in the amniotic fluid and membranes of the mother's original offspring. Stimulating the cervix of the ewe that is to adopt the offspring as it is presented can also enhance the process. This fools the mother into believing that the newly introduced offspring is her own. Once the mother and her new offspring have established their strong bond, the mother will rear the young animal as if it were her own (see Fig. 10.3). Cross-fostering is a useful research tool in that it allows scientists to distinguish between behaviours that are innate (i.e. that are genetically inherited) and those that are learnt by the offspring from the mother. Fostering is an important commercial husbandry procedure as it allows the farmer to get ewes (or nanny goats) that have lost their own offspring to adopt orphaned lambs (or kids).

Development of the mother–infant bond

The strong bond that develops between the mothers of both species and their offspring is initially based on olfactory cues, especially from the anal and genital region of the lamb. However, the mothers of both species can use other cues to identify their offspring. The importance of visual cues from various body regions of the lamb was demonstrated in a

Fig. 10.3. A nanny goat with twin lambs that have been cross-fostered to her.

study by Alexander and Shillito (1977). They used a black powder to blacken different parts of a lamb's body (see Fig. 10.4). They then tested the response of the lamb's mother when it confronted their partially blackened lamb for the first time. All of the ewes tested were hesitant or avoided their lamb as it approached when it had either a black head or when it was completely black. In contrast, lambs with a black rump or black foreheads were hardly avoided at all. The researchers concluded that the visual cues used by the ewe to identify her offspring come largely from the head of the lamb. Similar research has shown that ewes can also recognize the bleat of their lamb. The voices of goat kids are similar at birth, and the nanny goat cannot identify her offspring by sound alone until the sounds begin to diverge at about 4 days of age. Nanny goats and their kids typically remain isolated from the rest of the group for the first few days after birth.

As the lamb grows older and the mutual recognition improves, the lamb and ewe stray further apart. The ewe may call the lamb to suckle, but she also shows an increasing tendency to terminate suckling as the lamb grows older. Ewes with twins will only allow suckling when both lambs are present, indicating that she recognizes that she has two offspring. The frequency and duration of suckling gradually declines as the offspring grows older, until the mother suddenly becomes antagonistic when the young animal approaches to suckle. This results in the lamb or kid being weaned, although this is often enforced prematurely by the farmer in domestic animals.

Development of Lambs and Kids

Lambs and kids are born in a generally well-advanced stage of physical and behavioural development. Most lambs and kids are standing within 15 minutes of birth. As their coordination increases, the offspring nose

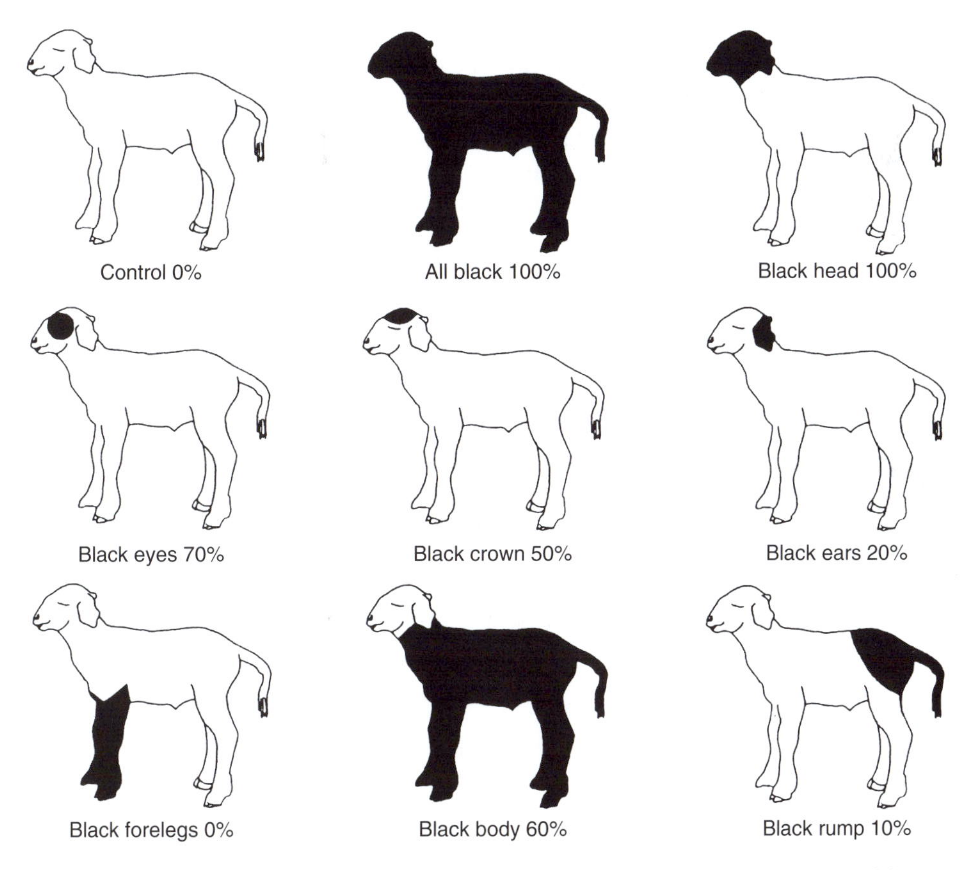

Fig. 10.4. The effect of blackening various parts of a lamb's body on the occurrence (shown as percentages) of hesitance or avoidance behaviour shown by the mother. (Adapted from Alexander, G. and Shillito, E.E., Importance of visual cues from various body regions in maternal recognition of the young in Merino sheep (*Ovis aries*), *Applied Animal Ethology* 13, 137–143. Copyright 1977, with permission from Elsevier Science.)

the side of the mother along the length of her body until they find a teat. Most lambs and kids will suckle within the first hour after birth. When suckling, the young will butt the udder, which facilitates milk letdown in the mother. The course of behavioural development in lambs and kids is, therefore, fairly rapid in the first day of life. Initially, lambs do not discriminate between ewes, and they will attempt to suckle from any ewe. However, lambs learn to use various cues to identify their own mothers as they grow older. By 3 weeks of age, visual cues have become more important to the lamb than auditory cues in identifying its own mother.

'Followers' versus 'hiders'

One of the main differences between sheep and goat behaviour is that lambs will follow their mother as she grazes (and are therefore described as being 'followers'). In contrast, goat kids (like calves) do not follow their mother, but remain hidden while she forages (known as 'lying-out'). The nanny goat leaves her kid for periods of between 1 and 8 hours, before returning to the precise location where the kid is lying. She makes a muted vibrato call, which signals the kid to rejoin its mother. Disturbance does not seem to frighten the kid while it is lying-out, although it may freeze if the mother makes an alarm call. This freezing response makes it more difficult for a predator to locate the kid. The 'lying-out' phase of the kid's development has been reported to last between 3 days and several weeks. After the lying-out period is over, the nanny goat and kid use bleating to keep in contact when moving or grazing.

Play behaviour

The young of both species show play activity from a young age. Lambs show clear sex differences in the types of play exhibited, with males engaging more in play fighting and females in more rotational–locomotor play. It is suggested that the locomotor play in the females ensures good physical fitness so that the animal is prepared to avoid potential predators.

Anti-predator Behaviour

Wild sheep show strong anti-predator behavioural responses. They are very shy and attentive, with a strong tendency to flock, and have quick flight reactions. However, sheep will defend their offspring when necessary, and will endeavour to chase away small carnivores. Bighorn (*Ovis canadensis*) rams will even form a defensive 'musk ring' around the flock to keep predators at bay. Although centuries of domestication by humans have reduced these anti-predator responses, domestic sheep still show many of these original behaviours. In contrast to goats, modern domestic sheep will crowd together when approached by a potential predator, making them relatively easy to herd as a group, especially using a trained sheep dog. Sheep show a stronger anti-predator response to models of carnivores than to neutral stimuli. For example, sheep took longer to recover and showed a greater flight distance when exposed to a stuffed wolverine, lynx or bear on a trolley, or a man in a poncho with a dog, compared with a ball on the trolley, the trolley alone or a man in a poncho without a dog (Hansen *et al.*, 2001). Lighter-bodied (typically less domesticated) sheep breeds show a stronger anti-predatory response than heavier breeds. The alarm posture in the sheep is to hold its head up rigidly while walking with quick, short steps. When alarmed, goats will stamp one foot and produce a high-pitched noise that sounds like a sneeze. These signals alert the other members of the flock to the potential threat. When approached by a potential predator (e.g. a dog), goats tend to break away from the flock. This makes it relatively difficult to 'herd' goats, as they do not group together.

Applied Problems

The welfare of sheep and goats

Although sheep and goats were amongst the first species to be domesticated, the farming methods used to keep them have generally not been subjected to the high level of intensification found with pigs, poultry or, to a lesser extent, dairy cattle. Sheep and goats are usually reared in extensive systems that are, in many respects, similar to the natural habitat of their wild ancestors. Consequently, sheep and goats (unlike pigs and poultry) do not typically suffer from behavioural problems associated with intensification. However, the fact that sheep and goats are kept in extensive conditions similar to those of their wild counterparts does not mean that their welfare cannot be compromised. Although animals in extensive systems can usually perform their full repertoire of natural behaviour, they can still be exposed to a variety of stressful situations. Animals kept outdoors can be exposed to extremes of heat and cold, and can suffer from poor nutrition if overstocked. Goats dislike rain, and it is important for the farmer to provide them with adequate shelters. When giving supplementary feed to sheep, it is important to ensure that sufficient feeding space is provided. If feed space is insufficient, less competitive sheep will simply stop feeding rather than fight for space.

Predation and fearfulness

Domestic sheep and goats are still exposed to carnivorous predators in many parts of the world. For example, in parts of Scandinavia, brown bear, wolf, lynx and wolverine all predate sheep. There are distinct breed differences in mortality losses attributed to predation, with the lighter breeds of sheep suffering less predation than the heavier breeds. This probably reflects differences in the anti-predator behaviour (discussed above), with the lighter breeds showing more pronounced anti-predator responses, and consequently suffering less predation. Although goats typically show less signs of fearfulness than sheep towards humans, rearing influences the response of goats to humans. Dam-reared goats show a greater avoidance of humans and exhibit greater impairment of milk ejection than human-reared goats. The impairment of milk ejection declines over time, suggesting that goats do eventually habituate to the aversive properties of the milking operation.

References

Alexander, G. and Shillito, E.E. (1977) Importance of visual cues from various body regions in maternal recognition of the young in Merino sheep (*Ovis aries*). *Applied Animal Ethology* 3, 137–143.

Clutton-Brock, J. (1999) *A Natural History of Domesticated Mammals.* Cambridge University Press, Cambridge.

Hafez, E.S.E. and Hafez, B. (2000) *Reproduction in Farm Animals*, 7th edn. Lippincott, Williams and Wilkins, Philadelphia.

Hansen, I., Christiansen, F., Hansen, H.S., Braastad, B. and Bakken, M. (2001) Variation in behavioural responses of ewes towards predator-related stimuli. *Applied Animal Behaviour Science* 70, 227–237.

Lynch, J.J., Hinch, G. and Adams, D.B. (1992) *The Behaviour of Sheep: Biological Principles and Implications for Production.* CAB International, Wallingford, UK.

Mowlem, A. (1996) *Goat Farming*, 2nd edn. Farming Press Books, Ipswich, UK.

Parsons, A.J., Newman, J.A., Penning, P.D., Harvey, A. and Orr, R.J. (1994) Diet preference of sheep: effects of recent diet, physiological state and species abundance. *Journal of Animal Ecology* 63, 465–478.

Rutter, S.M., Orr, R.J. and Rook, A.J. (2000) Dietary preference for grass and white clover in sheep and cattle: an overview. In: Rook, A.J. and Penning, P.D. (eds) *Grazing Management. BGS Occasional Symposium No. 34.* British Grassland Society, Reading, UK, pp. 73–78.

11 Behaviour of Pigs

Per Jensen

The Origin of Pigs

Pigs originate from the European wild boar (*Sus scrofa*). Based on archaeological findings, domestication is estimated to have started about 10,000 years ago (Clutton-Brock, 1999). One problem for domestication researchers has been the fact that the wild boar is an excessively variable species (16 different subspecies have been proposed), and even Darwin noted that domestic pigs seem to come in two distinct forms – one European and one Asian. Studies at the level of molecular genetics lends strong support to the theory that pigs were independently domesticated from wild boar subspecies in Europe and Asia, and also suggest that reproductive isolation between present domestic and wild populations occurred long before archaeological findings started to emerge (Giuffra *et al.*, 2000).

As the human species started to expand and spread, about 100,000 years ago, wild boars were common. It is not unlikely that populations of wild boar, both in Asia and Europe, associated with humans early on, and perhaps took advantage of this cohabitation. As agriculture emerged, these populations may have been particularly suitable for active domestication and breeding (see also Chapter 2). The earliest use of pigs was obviously for food.

As we will explore the normal behaviour of domestic pigs in this chapter, we will make use of three different sources of information. First, the behaviour of the ancestors, wild boars, may provide insight into the functions of different behaviours. Secondly, the behaviour of feral pigs (populations of domestic pigs which have escaped captivity and reproduce freely with little influence from humans) may tell us something about how well conserved the behaviour of the wild boars has been through domestication. Thirdly, researchers have released domestic pigs into large, natural enclosures, and studied their behaviour under more-or-less undisturbed conditions. Such studies also provide detailed knowledge of the behaviour that has been retained unchanged throughout generations of life under human control.

Social Behaviour

Pigs are generally social animals. In wild boar and feral pigs, the typical pig herd is made up of closely related females and their offspring, whereas sexually mature boars often are solitary, or sometimes live in all-male groups. The group sizes of the female herds are about 2–6 individuals (Graves, 1984).

When pigs are released into nature, they tend to form the same types of groups (Jensen, 1988). Long-term relations are already established between litter-mates early in life, and these social relations are obvious long thereafter.

A pig group develops a stable hierarchy, which to a large extent is maintained by active submission and avoidance behaviour by the animals which are low in the social order. When pigs under production conditions are mixed with strangers, they normally fight intensely for a period until a dominance order has emerged. These fights are almost always won by the bigger individuals, and consequently there is normally a strong correlation between size and dominance status in a group (Jensen, 1994).

Wild boars and feral pigs are not territorial, i.e. they do not defend a specific area against conspecifics, but they live in restricted home ranges and show a high degree of site fidelity (Graves, 1984). The home range sizes can vary widely; different studies have reported from less than 100 to more than 2500 ha. Food availability is important in determining home range size, and males often have larger ranges than females. Domestic pigs in natural enclosures use their available space extensively, although there are no data on actual home range sizes. Apparently, food availability is important to domestic pigs as well, and in addition, the ranges decrease in size during winters (Jensen, 1988).

Signals and Communication

The most important sense for social interactions in pigs is olfaction. Chemical signals are used for individual recognition and there are probably odours (pheromones) that carry very specific messages. For example, a submissive pheromone has been implicated by some experimental studies (McGlone *et al.*, 1987).

Sight seems to be less important to pig communication. Pigs fitted with opaque eye-cups are basically unaffected in their social behaviour (as opposed to pigs made anosmic, where social relations break down immediately). Nevertheless, specific visual signals have been described in wild boars and domestic pigs. Both ears, tails and body posture are used for signalling. Ears that are held back along the neck signal fearfulness, tail erect and upwards signals danger, whereas a depressed tail is typical of a submissive pig. Tilting the head to the side is a submissive signal, whereas arching of the back could be a threat.

It is a common misconception that the curled tail of a domestic pig signals 'happiness'. Wild boars do not curl their tails – apparently, this is an effect of domestication, analogous to the curled tails of some breeds of dogs. The idea has some validity in that a relaxed pig, which is not exposed to danger or social threat, tends to curl the tail, but this does not seem to carry any signalling function to other pigs.

Pigs also possess a repertoire of vocal signals, with more-or-less specific functions. For example, the warning call, which is emitted as a response to a sudden frightening stimulus, is a dog-like bark. Other pigs immediately respond by repeating the call and either freeze or run away. There are also contact grunts, submissive squeals, begging screams of piglets and the special lactation grunt of the sow, which will be described in some detail later in this chapter (Kiley, 1972).

Foraging and Feeding

Pigs are omnivorous animals and readily adapt their diets within wide limits to the prevailing conditions. In wild boars and in feral pigs, the basis of the diet is usually plants and plant-based food items, such as grass, roots, fruit, berries, seeds, etc. Animals may also constitute a substantial part of the diet. Remains of earthworms, frogs and rodents have been found in the stomach contents of wild boars. Pigs may even act as predators. In parts of the world (e.g. Australia and New Zealand), feral pigs are considered to be pests, since they may adopt the habit of attacking, killing and eating new-born lambs.

The snout of the pig is extremely well adapted to its feeding habits. With its upper part, it can lift heavy objects such as stones and logs, and dig and turn the soil over to get access to roots and seeds (Fig. 11.1). The

Fig. 11.1. Pigs in natural conditions live in groups. One important way of foraging is to root, whereby the soil is turned over as the pigs seek roots and other food items.

disc of the snout is sensitive and well innervated, and the olfactory system is extremely sensitive. However, rooting is only one of the pig's typical feeding behaviours. Grazing and browsing is used extensively as well, in particular during the time of the year when green plants are abundant.

Is rooting a need in pigs? If allowed the possibility, most pigs will use a large portion of their time to go through rooting motions, even if only on a concrete floor. Sows on pasture are often equipped with nose rings to prevent them from rooting and destroying the land (since the rings cause pain if they attempt to root). If they do not have nose rings, they root extensively. Ringed sows also show some evidence of frustration, and all together these findings indicate that rooting may constitute a need in its own right in pigs (Horrell *et al.*, 2001).

Exploration is closely linked to foraging behaviour in omnivorous species, and pigs are no exception. Rooting, smelling and chewing are principal exploratory behaviours. Such investigative behaviour develops early under natural conditions and fills a substantial part of the time budget of free-ranging pigs. Even when there are no novel external stimuli, pigs appear to be strongly motivated to carry out exploratory behaviour (Wood-Gush and Vestergaard, 1993).

Diurnal Rhythms

Pigs are normally diurnal, but seem to be able to shift to nocturnal activity rather easily. Wild boars are often nocturnal in their activity cycles, in particular in regions where they are hunted.

Under housing conditions, it is common that the activity periods are centred around feeding times. This may cause a typical diurnal rhythm of, for example, two activity peaks, one in the morning and one in the afternoon, interspersed by resting periods. Under free-range conditions, weather is an important determinant of activity. Under hot conditions, and in very harsh weather, pigs usually rest during the day and become active at dusk and dawn. Since pigs depend on behavioural thermoregulation to a large extent (wallowing when hot, huddling when cold), this affects their diurnal activities as well.

Wallowing deserves a special mention, since this is a commonly known pig behaviour (Fig. 11.2). Pigs have very few sweat glands, and are almost incapable of panting. Instead, they rely on wallowing in water or mud to cool the body when the ambient temperature rises. Adult pigs under natural conditions can often be seen to wallow when temperatures exceed 20°C. Mud is the preferred substrate, and after wallowing the wet mud will provide a cooling, and probably protecting, layer on the body for an extended time. When pigs enter a wallow, they normally dig and root in the mud before entering with the forebody first. They then wriggle the body back and forth so all of the body surface is covered with mud, before shaking the head and often finishing with rubbing against a tree or a stone next to the wallow. Indoors, pigs may attempt to wallow on wet floor surfaces and in the dunging areas when it is hot.

Under free-range conditions, domestic pigs move between specific foraging areas and use a few specific resting sites day after day. More-or-less permanent resting sites are also known from feral pigs and wild boars, and – as in free-ranging domestic pigs – these sites are often formed into resting nests, by means of the pigs bedding with grass, leaves and twigs. Such resting nests are also used for the longer night sleeping periods. Normally, the whole group stays in the same nest, huddling close together. Improvement of the resting nests by carrying out nest-building activities can be seen prior to each resting period. In indoor pens, pigs will often root and paw on the floor and carry straw (if available) to the resting place before lying down for the night.

Mating Behaviour

Domestic pigs become sexually mature earlier than wild boars – at about 7 months of age compared to 1.5 years. In Europe, wild boars have their peak mating season in the autumn, but matings and farrowings can occur at any time of the year. When there is plenty of food and suitable habitats, wild sows may give birth to two litters a year.

This seasonal plasticity has been an important base for the success of using pigs for food production. Domestic pigs will come into heat at any time of the year, even if there are some signs of a remaining annual cyclicity. When not pregnant, sows will come into heat at 21-day intervals. Each heat period lasts 3 days, with a maximal receptivity period of about 12 hours.

Fig. 11.2. Wallowing.

Different experiments have shown that boars are rather indiscriminative to different sows. When sexually aroused, they do not even appear to distinguish a sow in heat from one in anoestrous when given a choice. Sows, however, show a definite discrimination during oestrous (but not during anoestrous), preferring intact males over castrated ones, and mature boars over young ones (de Jonge *et al.*, 1994).

The courting and mating behaviour of pigs contains arrays of signals from both parties. The sow urinates and squeals as the boar courts her. The boar consistently emits so-called staccato-grunts (also referred to as 'chant-de-coeur'), urinates and produces foaming saliva. He also pushes the sow's side with the snout and places the head repeatedly on her back (Fig. 11.3). When the sow is receptive, she assumes a specific body position, called 'standing posture'. This elicits the mounting. The penis of the boar is spiral shaped, corresponding to the shape of the cervix, which allows the semen to be placed high up in the reproductive tract. The copulation is unusually long compared to other ungulates, and may last 5–10 minutes. This, combined with the large volume of semen in each ejaculate, may be interpreted as a sign of sperm competition – since different piglets in the same litter may have different fathers, there has been evolutionary pressure on boars to increase the chances of their own sperm reaching the ova.

Fig. 11.3. During courtship, the boar stimulates the sow by pushing at her side, grunting and emitting pheromones.

Nest Building and Parturition

Following a pregnancy that lasts about 115 days, sows in free range usually leave the group and the central home range and seek out a suitable farrowing site. After leaving, the sow may wander distances of many kilometres before settling for a place (Jensen, 1988). Typical features of preferred farrowing sites are protection towards one side, in the form of a slope or a stone, and vertical protection in the form of overhanging branches.

As soon as the site has been chosen, the sow starts nest building. Interestingly, pigs are the only ungulates to perform nesting behaviour. Nest building follows a rather fixed pattern, starting with the sow rooting and digging a shallow hollow in the ground. Thereafter, soft material is ripped off from the edges of the hollow and, with pawing movements of the forelegs, the material is placed in the nest. The next phase consists of the sow collecting nest material – grass, ferns and twigs – from distances up to 50 m from the nest site, carrying it back and placing and arranging it in the nest. Arranging is done by rooting and pawing in the material so it becomes accumulated along the edges of the nest and gradually piles up in the middle.

When the nest is finished, the sow enters it by kneeling and pushing herself forwards underneath the nest material. If the sow has had access to suitable material, she may become completely covered by nest material after entering the nest.

Experiments have shown that sows will attempt to perform nest building in any environment. It is largely triggered by prostaglandins and may actually be released by injection of such substances in non-pregnant animals (Gilbert *et al.*, 2000). It appears that proper nest-building activities have a priming effect on subsequent maternal behaviour, since sows that have been deprived of the possibility to nest have been observed to have poorer maternal responsivity.

Parturition follows within a few hours after the nest building is finished. Unlike other ungulates (or indeed, most other mammals), the sow does not lick the new-born piglets and rarely gets up to sniff at them. Each of the 10 or so piglets frees itself of the fetal membranes and tears the umbilical cord in its attempts to get to the udder. Under free-range conditions, farrowing often lasts 4–6 hours. In production units, it may take several hours longer.

Suckling

In free-ranging pigs, the sow remains with the piglets in the farrowing nest for a period of up to 2 weeks – an average of 10 days has been reported (Jensen, 1988). During this period, more or less all suckling takes place in the nest, even though the piglets start following the sow on excursions after a few days.

During the first days of life, a teat order develops within the litter. Each piglet attaches to one, or sometimes two, teats which they consistently use thereafter. In the first weeks after parturition, sows nurse about once per hour, and each suckling contains a complex signalling pattern between mother and young (Fig. 11.4).

The sow always lies down to nurse during the first weeks, and signals the start of the nursing by emitting a deep pitched, distinct lactation grunt. This grunt is repeated in about 2-second intervals over approximately 1 minute. The piglets gather at the udder, find their own teats and start massaging the udder by up-and-down movements with the snouts. After a minute or so, the rate of grunting increases to about two grunts per second, which causes the piglets to stop massaging and start to suck slowly on the teat. Another 20–25 seconds after the peak of the grunt rate, milk flow starts. The total milk ejection time is only

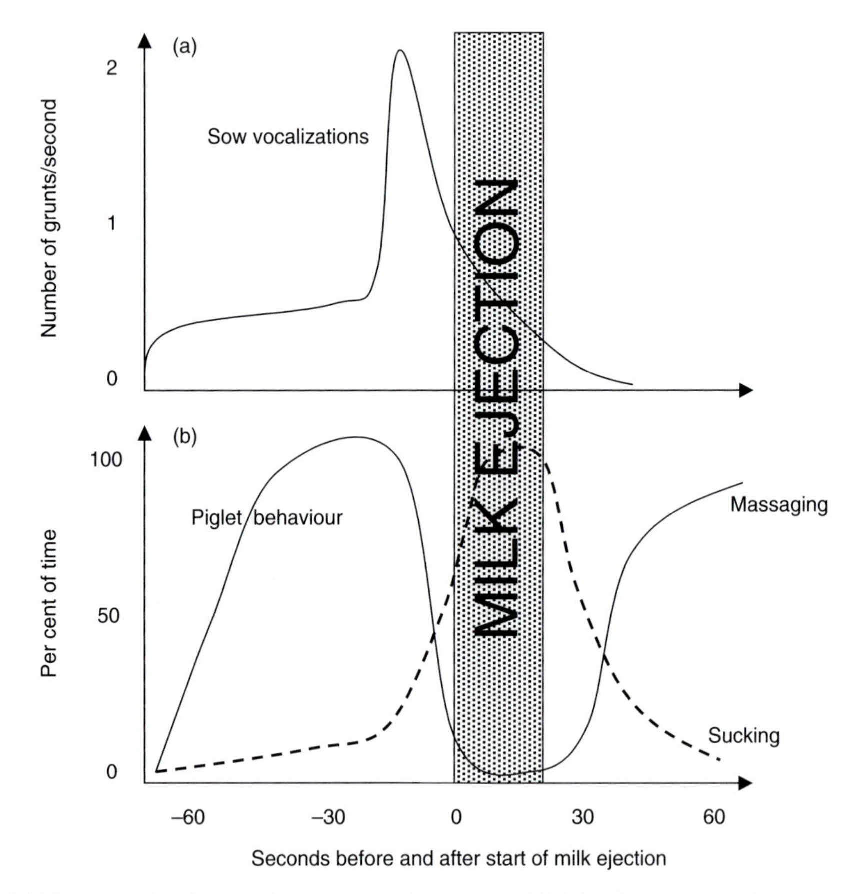

Fig. 11.4. (a) The rate of emission of grunts from the sow, and (b) the time pattern of two behaviours of the piglets – massaging and sucking on the teat. Typical, average values are shown. (Derived from Fraser, 1980.)

about 20 seconds, and thereafter, no more milk can be extracted. The piglets resume massaging and may continue with this for 10–15 minutes (Fraser, 1980). The so-called post-massage behaviour appears to be a way for each piglet to stimulate further milk production in the teat that it uses, sometimes at the expense of milk production in the teats of the siblings (Jensen *et al.*, 1998).

Development of the Young

Piglets are very precocial in their behaviour, and move with well-coordinated movements within less than an hour after birth. In fact, this is an exceptional mammalian neonate behaviour. As a general rule, offspring which are born in large litters are altricial, and often blind and deaf, like rodents and many carnivores. Offspring that are born in small litters, or as single offspring, are normally precocial, as in other ungulates. Pigs are one of very few mammals to have precocial young born in large litters.

After a couple of days, the piglets in free-range conditions begin to follow the sow when she leaves the nest to feed and drink, and, as mentioned above, after about 10 days, the sow brings them to the rest of the herd. From that time, they usually never return to the farrowing nest.

This is usually the first time a piglet will encounter non-family members, and one might expect that there would be a lot of fighting. However, even though the social activity is high during some weeks after nest leaving, there is little overt aggression. Most of the interactions consist of nose contacts, and when the piglets are 8 weeks old, the interaction frequency has stabilized at a low level. The period between 2 and 8 weeks of age has therefore been called the phase of social integration (Jensen, 1988).

Under natural conditions, piglets start to ingest some solid food immediately after birth, but only when they are about 5 weeks old do they start to eat substantial amounts. Also after that, they continue to suckle for an extended period. The average weaning age (the age when suckling is observed for the last time) was about 17 weeks in one large study (Jensen, 1988). Weaning is a slow and gradual process, where suckling frequencies decrease almost linearly from soon after birth up to the final milk intake (Fig. 11.5).

Applied Problems

Modern pig housing is far from the natural situation in enclosures or the wild. This causes concern among scientists and animal welfare organizations alike. Although the welfare is not necessarily optimal, or even good, in nature, there are always large risks in forcing animals to live under conditions where their natural behaviour is severely constrained.

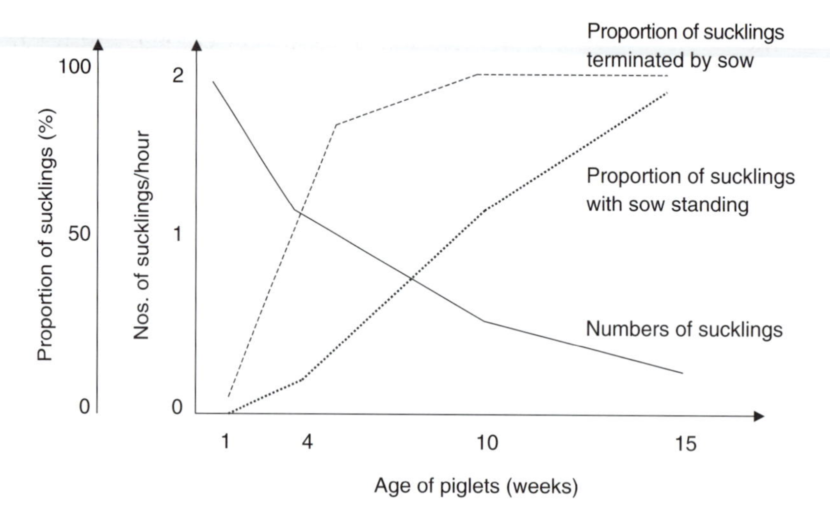

Fig. 11.5. During the last period of weaning, the piglets initiate most sucklings, and the sows often attempt to get away from the piglets, and terminate most sucklings. The diagram shows some typical average figures of numbers of sucklings per hour, proportion of sucklings terminated by the sows, and proportion of sucklings performed while the sow is standing up, at different ages of the offspring in free-range conditions. (Adapted from Jensen, 1988.)

Space restrictions

One of the most obvious potential problems in pig housing is the lack of space. In large parts of the world, sows are kept confined or tethered in very small areas. Fattening pigs are crowded in pens where they may not even have space to lie down on their sides all at once. In Europe, pregnant sows are more and more often kept in groups, and many countries, including the EU, have enforced this bylaw. However, this does not include farrowing and lactating sows, which are still kept confined in most countries.

Lack of stimulation

An even more substantial problem than space alone is the structure of the space. Pigs are active and exploratory, but the pens are often barren and without stimuli or structures for their natural behaviour. For example, a farrowing sow can rarely perform nest building, which may be a considerable source of poor welfare (Lawrence *et al.*, 1994). Floors are often slatted or consist of wire or plastic netting, where it is not possible to use straw. Such environments are lacking in stimuli and do not allow for much exploration.

A somewhat related problem is the food restriction that is normally inflicted on pregnant sows (and often on older fattening pigs). Since pregnant sows cannot be allowed to grow too fat, which may cause problems at farrowing, food is usually restricted to a level well below that which the sow would eat with free access. Consequently, sows are usually hungry for most of the pregnancy, and this has been found to be an important cause of stereotypies.

Surgery and mutilation

Some surgery, or mutilations, are routinely performed on pigs in order to prevent certain problems. In new-born piglets, the sharp canine teeth and third incisors are often cut or ground down. This is because these teeth may cause injuries to siblings, in particular if the litters are large, and sometimes also to the udder of the sow. However, the cutting may cause pain and lead to infection if not done absolutely correctly, and it may be a better strategy to cut teeth only in large litters (say, more than 12 piglets) or to cut selectively only the teeth of the largest piglets, in order to leave some competitive advantages for the smallest. However, selective cutting appears not to have very profound influence on the survival and growth of the smaller piglets.

Male piglets are castrated during lactation, often during the first 1–2 weeks, in most countries, because the meat of uncastrated males may develop an unpleasant smell and taste, so-called boar taint. Castration is normally done without anaesthesia, and several studies have demonstrated that the reactions of the piglets clearly indicate that this is a painful procedure, and that the piglets may continue to be in pain for 1–2 days. Contrary to common belief, the reactions of the piglets do not seem to be less severe when they are very young (1 week).

Also, in most countries, the tails are cut off the young piglets, again without anaesthesia, to prevent later development of tail-biting behaviour. This is a potential behavioural problem, in that the tail has important signal functions in the communication between pigs. It is also a general welfare problem. Some research has found that neuromas may develop in the cutting area of the tail, and such neuromas in humans are associated with so-called phantom pain.

Weaning

Another aspect of pig husbandry where there are substantial deviations from the natural behaviour is weaning. First, weaning is usually carried out much earlier than in nature, often when the piglets are less than 4 weeks old. Secondly, weaning in pig production is mostly abrupt – the mother is instantaneously removed from the litter. Thirdly, it is common in many countries that piglets are transferred to weaning pens which are very barren and crowded. Fourthly, it is not uncommon that weaning is associated with mixing of litters. All these events are potentially stressful for the piglets and may cause various problems, such as increased disease susceptibility (for example, diarrhoea) and behaviour disturbances.

Mixing of unacquainted pigs

Mixing of pigs that are not familiar to each other is always a source of problems, since it will inevitably lead to severe fighting. Many methods have been tried, both scientifically and practically, to reduce fighting after mixing, including application of strong odours or medication with sedatives. However, no single method has proven successful. Based on various experiments and on what we know about the normal behaviour of pigs, it is possible to set out a few recommendations. First, if mixing should be done, it is best to mix piglets when they are very young, preferably 2–3 weeks old. Secondly, it is better to mix piglets when the mothers are present. Thirdly, it is better to mix any group of pigs in a complex environment, containing, for example, lots of straw and other stimuli, including hiding areas.

Abnormal behaviour

Confined pregnant sows often develop stereotypies such as bar biting and sham chewing. Partly, this can be explained by a low feeding intensity and a frustrated motivation to forage, but many studies have shown that stereotypies are generally more common in confined than in loose sows. As mentioned in Chapter 6, there are different scientific interpretations of the significance of stereotypies, but it is mostly agreed that a high level of such behaviour is a clear sign of reduced welfare of the animals (Lawrence and Rushen, 1993).

In younger pigs, specifically in early weaned piglets and in fattening pigs housed in crowded and barren conditions, belly nosing and tail biting may occur. Belly nosing is a behaviour where one or several pigs perform rooting snout movements towards the belly of a pen-mate. The behaviour is often interpreted as a sign of a strong motivation to perform suckling activities, and it may be harmful through causing injuries to the recipient. Tail biting (and the closely related ear biting) is a behaviour that appears most common in crowded and stimulus-poor environments. It is mostly believed not to be an aggressive behaviour, but rather originates from a strong motivation for exploring. When chewing and manipulating of the tail leads to blood flow, this may in turn trigger massive attacks on the bleeding pig, which may even lead to the death of the exposed pig. Again, experimental data suggest that this is mainly caused by an urge to explore the new stimuli arising from the blood, and does not seem to be motivated by any aggression.

Ethological Pig Housing

Is it necessary to keep pigs under completely unnatural conditions in order to have efficient pig production? The answer is no. In many countries, a large proportion of the sows are kept on outdoor pastures, both when pregnant and during farrowing. However, these animals are often nose ringed to stop them from rooting, and this of course is a serious limitation to their natural behaviour.

In Sweden and some other European countries, there are systems in practical pig production where both pregnant and lactating sows are kept loose in groups. In some systems, the sows may themselves select a nest site (out of several small pens provided by the farmer) and build a nest out of straw (Fig. 11.6). When the piglets reach about 2 weeks, they can leave the nest and mix with other litters. Some farmers run this system with very high production results.

In the case of fattening pigs, fewer well-designed alternatives exist. However, there are systems where the pigs are kept in large pens with plenty of straw and structures to make the environment more complex. This usually means that group sizes are much larger than normal, so the average available area per pig does not increase very much. There is clearly a need for more research on fattening pig housing systems which allow a more natural behaviour without compromising other aspects of pig welfare or farm economics.

Fig. 11.6. One example of a group-housing system for lactating sows, the so-called Thorstensson system, where the sows farrow in small, detachable farrowing pens, which are removed after about 2 weeks. The sows can move in and out of the farrowing pens all the time, but the piglets are kept inside by a step.

References

Clutton-Brock, J. (1999) *A Natural History of Domesticated Mammals.* Cambridge University Press, Cambridge.

de Jonge, F.H., Mekking, P., Abbott, K. and Wiepkema, P.R. (1994) Proceptive and receptive aspects of oestrus behaviour in gilts. *Behavioural Processes* 31, 157–166.

Fraser, D. (1980) A review of the behavioural mechanism of milk ejection of the domestic pig. *Applied Animal Ethology* 6, 247–255.

Gilbert, C.L., Murfitt, P.J.E., Boulton, M.I., Pain, J. and Burne, T.H.J. (2000) Effects of prostaglandin $F_{2\alpha}$-treatment on the behaviour of pseudopregnant pigs held in an extensive environment. *Hormones and Behaviour* 37, 229–236.

Giuffra, E., Kijas, J.M.H., Amarger, V., Carlborg, Ö., Jeon, J.-T. and Andersson, L. (2000) The origin of the domestic pig: independent domestication and subsequent introgression. *Genetics* 154, 1785–1791.

Graves, H.B. (1984) Behaviour and ecology of wild and feral swine (*Sus scrofa*). *Journal of Animal Science* 58, 482–492.

Horrell, R.I., A'Ness, P.J.A., Edwards, S.A. and Eddison, J.C. (2001) The use of nose-rings in pigs: consequences for rooting, other functional activities, and welfare. *Animal Welfare* 10, 3–22.

Jensen, P. (1988) *Maternal Behaviour of Free-ranging Domestic Pigs. I: Results of a Three-year Study.* Swedish University of Agricultural Sciences, Department of Animal Hygiene, Report 22, Skara.

Jensen, P. (1994) Fighting between unacquainted pigs: effects of age and of individual reaction pattern. *Applied Animal Behaviour Science* 41, 37–52.

Jensen, P., Gustafsson, M. and Augustsson, H. (1998) Teat massage after milk ingestion in domestic piglets: an example of honest begging? *Animal Behaviour* 55, 787–797.

Kiley, M. (1972) The vocalisations of ungulates, their causation and function. *Zeitschrift für Tierpsychologie* 31, 171–222.

Lawrence, A.B. and Rushen, J. (eds) (1993) *Stereotypic Animal Behaviour – Fundamentals and Applications to Welfare.* CAB International, Wallingford, UK.

Lawrence, A.B., Petherick, J.C., McLean, K.A., Deans, L.A., Chirnside, J., Vaughan, A., Clutton, E. and Terlouw, E.M.C. (1994) The effect of environment on behaviour, plasma cortisol and prolactin in parturient sows. *Applied Animal Behaviour Science* 39, 313–330.

McGlone, J.J., Curtis, S.E. and Banks, E.M. (1987) Evidence for aggression-modulating pheromones in prepuberal pigs. *Behavioural Neural Biology* 47, 27–39.

Wood-Gush, D.G.M. and Vestergaard, K. (1993) Inquisitive exploration in pigs. *Animal Behaviour* 45, 185–187.

12 Behaviour of Dogs and Cats

Bjarne O. Braastad and Morten Bakken

Origin and Domestication History

Dogs

The modern-day domestic dog (*Canis familiaris*) is one of the 38 species within the canine family, Canidae. These are a biologically cohesive group of predators which diverged developmentally from other predators about 10 million years ago. The family includes various species of fox (*Vulpes vulpes*, *Alopex lagopus*), wolf (*Canis lupus*), coyote (*Canis latrans*), African wild dog (*Lycaeon pictus*) and several species of jackal, e.g. the golden jackal (*Canis aureus*).

Whether domestic dogs originated from only one, or from several different wild species has been discussed for several hundred years. The most likely candidates were wolves and the golden jackal. Both species have the same number of chromosome pairs as domestic dogs (39), have similar behaviour and can produce fertile offspring when mated with domestic dogs. Using modern biotechnological methods combined with a knowledge of behaviour, morphology and vocalizations, the currently accepted theory is that all domestic dogs originated from diverse subspecies of wolf: the northern grey wolf, the small desert wolf of Arabia, the pale-footed Sian wolf and the Chinese wolf.

Wolf remains from over 100,000 years ago have been discovered alongside remains of early hominids. Human predecessors may have hunted wolves as a nutritional source, or to utilize fur for clothing. It is possible that individual cubs were later caught and raised by *Homo sapiens*; those that adapted remained in the human environment. These individuals may have been predecessors to animals that, through countless generations, developed into domestic dogs. Archaeological findings from approximately 14,000 years ago reveal remains that differ from wolves, being smaller and having a shorter snout – the first domestic dog. As the exact nature of the evolutionary history of domestic dogs is a mystery, so is the exact nature of the relationship between humans and dogs in the early days of domestication. Humans were still hunters, with no weapons other than stones and clubs. One theory claims that humans became more effective hunters because of dogs. Others maintain that dogs' ability to warn against danger formed the basis of the coexistence between humans and dogs. In addition, the presence of dogs may have kept wolves and other predators at a distance, and kept sites free of food remains, waste and human excrement.

Archaeological findings reveal great diversity in both body size and proportions between prehistoric dogs from different geographical areas. The first main types of the modern-day dog were seen in drawings, frescoes and art 3000–4000 years ago. Large dogs of the mastiff type in Asia were found in Babylonian frescoes, greyhounds were reflected in art from Egypt and dogs of the Pekinese type were reflected in art from China. A record of breed development is also available through historical literature. The Romans had diverse names for dogs according to their purpose; house dogs, hunting dogs, war dogs, sheep dogs, tracking dogs and dogs that hunted using visual skills. The Romans also realized that it was possible to alter dogs' appearance and behaviour through specific breeding. Through these archaic dogs, humans have bred over 400 different dog breeds, according to their requirements for specialized assistance in hunting, guarding, mustering, but also for their fighting abilities and attractive appearance (Morey, 1994; Clutton-Brock, 1999).

Cats

The domestic cat (*Felis catus*) belongs to a group of small cat species around the Mediterranean region that diverged from the rest of the cat family, the Felidae, around 8–10 million years ago. The domestic cat is most similar to the European wildcat (*Felis silvestris*) and the African wildcat (*Felis libyca*). It is now generally thought that these all belong to one polytypic species, called *Felis silvestris*. While it is difficult, on an anatomical basis, to reveal which of these is the most probable ancestor, both archaeological evidence from about 6000 BC and recent behavioural evidence favour the African wildcat. In contrast to the extreme fierceness and resistance to taming of the European wildcat, the African wildcat is rather docile, lives closer to human settlements, and is more easy to tame. It is suggested that the oriental cat breeds have been derived from the Indian desert cat, *Felis silvestris ornata*.

The domestication of the cat is indicated by the presence of cat bones in a man's grave from 4000 BC and by tomb paintings in ancient Egypt about 2000 BC. The cat is sometimes depicted as hunting rats in human homes. The animal-loving Egyptians probably kept cats both as protectors of food stores and as pets. Cats were even associated with several goddesses, particularly those characterized by sexual energy. The best known of these, Bastet from the south-eastern part of the Nile delta, was depicted with a lioness head. Gradually the domestic cat came to be regarded as a manifestation of Bastet, giving the cat a protected status.

By this time, the domestic cat had already spread around the Mediterranean. Later, cat-keeping spread all over Europe, and, for example, was introduced to Scandinavia by the Vikings. In contrast to its high status elsewhere, in medieval England the cat was regarded as a demonic companion of witches. In recent years, however, the domestic cat has taken over the dog's role as the most common companion animal (Turner and Bateson, 2000, Chapter 9).

Social Behaviour

Dogs

The social behaviour of dogs and wolves resembles that of many other large canids. Canids are usually monogamous, reproduce once a year, and give birth in caves or dens. They have large litters compared with most mammalian species. Offspring are small and helpless at birth. Canids may exist in social groups based primarily on kinship, where family members assist in territory defence and care of the young. The larger canid species have effective food distribution to the young, by regurgitating food for the cubs.

Canids have a flexible social organization and the basic pack structure will depend primarily on the availability of food and type of prey within their niche. The largest pack sizes appear in coyotes and wolves, where the main prey is large ungulates such as elk and moose. However, in areas where the main prey is smaller ungulates and rabbits, pack size is significantly reduced, and may vary from seven to two individuals. Studies of feral dogs also reveal a parallel pattern – living in large packs in rural areas with abundant nutritional sources, and smaller packs in urban areas with poorer access to food. Canids in urban areas usually live in packs of only two or three individuals.

The dominant pair (alpha-pair) are usually the only individuals to reproduce in the social group. The so-called helpers are often related to the alpha-pair, most commonly cubs from previous litters. Research concerning the factors determining which individuals will be helpers and which will leave the social group is sparse in free-roaming canids. However, it is possible that population density, and availability of food and territory are decisive factors (Gittleman, 1989).

Cats

The cat's traditional reputation for being a solitary species is by far an oversimplification. Much of the cat's behaviour is dedicated to its relation with other individuals. The cat shows a flexible variation in its social behaviour, from solitarity to living in large groups, and a range of population densities from less than 1 to more than 2000 per square kilometre. Average home ranges vary between 0.3 and 170 ha in females and between 0.7 and 620 ha in (intact) males, largest in dominant males. Home ranges may overlap considerably between males, but not between colonies of group-living females. The remarkable variation in density and home ranges may be partly related to how predictable and patchy the important resources are for the cat.

When living in groups, adult females and their offspring may form core groups. These may be territorial. Several core groups may occupy high-quality sites and form a colony. Some competition for resources within the female colony may occur, giving rise to two social classes, 'central females' and 'peripheral females'. The central ones have a higher reproductive success and stay healthier than the peripheral ones. Adult males may be associated with such colonies. These may be called 'central males', in contrast to the 'peripheral males' that roam widely. This social structure is astonishingly similar to that of lions.

Studies of free-living and laboratory colonies of cats indicate that an absolute hierarchy in which dominant individuals have priority to resources is more pronounced in high-density groups. When population density is low in relation to available resources, cats are frequently not very territorial. Territoriality is most pronounced in solitary, reproducing females. Neutered males stay closer to their core area, in which they may be more territorial than males usually are.

The latter phenomenon may be utilized when dealing with problems of large colonies of feral cats in cities. Earlier attempts to kill such cats in large numbers only resulted in immigration of peripheral or migrating cats. Now a new method has been developed: cats are trapped, tested, vaccinated, altered (neutered or spayed), and subsequently released if their health is acceptable (the TTVAR method). Due to the territoriality of neutered cats and their inability to reproduce, population density is kept reasonably low. The long-term effect of this method is still uncertain, and there is a need for studies on how frequently such interventions should be performed (Leyhausen, 1979; Turner and Bateson, 2000, Chapter 6).

Communication

The complex social systems of canids and felids rely on effective communication between individuals for their maintenance and the individual's ability to separate group members from unfamiliar individuals. This complex communication is achieved via a combination of various methods: olfaction, vision, auditory means and physical contact. Although most communication signals are used for interactions with conspecifics, some are more frequently seen in interaction with humans. This may be particularly so in domestic cats, as it has been pointed out that in wild subspecies of *Felis silvestris*, for example, adults rarely use miaows. The individuality of these sounds indicates that they are at least partly modified through interactions with humans.

Dogs

Visual communication

Wolves and dogs have many similar visual postures, particularly in relation to dominance, aggression and fear. The individual's posture, ear position, lips, degree of eye opening and tail position all reflect the emotional state of the animal. For example, erect ears reveal alertness, while ears pressed against the head reveal submission or fear. A wagging tail indicates excitement, an upright tail indicates alertness, a low tail exhibits submission, and the tail between the legs shows fear. Lips drawn backwards indicate a threat. A dominant dog will have an upright posture, hold its head high and its ears erect, while a subordinate individual will have a lowered head, tail and body posture, with its ears laid flat. When aggression increases, both dominant and subordinate individuals will draw their lips back and show their teeth, and the hair across their shoulders will piloerect. However, the posture and ear position will vary according to the individual's position in the group. The dominant individual will make itself as large as possible and raise its tail, while a subordinate individual will have a low posture and tail position, with its ears pulled back.

The eyes also have a major communicative function: eye contact between a dominant and a subordinate wolf in a stable group will often suffice to cause the subordinate individual to lower its tail, lower its ears and retreat. Between dominant, unknown dogs eye contact may promote a fight.

When a subordinate individual seeks contact with a dominant individual, it lowers its body position, actively wags its lowered tail and nuzzles the lips of the dominant animal, in the same way as wolf cubs do when they stimulate their mother to regurgitate food.

Acoustic communication

Dogs have a vast repertoire of vocal signals such as barking, growling, whining, howling and grunting. Most vocal signals will have different meanings depending on the situation in which they are expressed. However, growls are used as a warning or threat signal and whining in defence or pain. Dogs appear to howl primarily when left alone – this can be construed as a signal for social contact.

Olfactory communication

A third channel of communication for wolves and dogs is communication based on chemical signals. These may be chemical components left behind from faeces and urine used by wolves to mark their territory. This is also often seen in dogs. In addition, general body odours produced by skin glands generally associated with the face, tail and anal region are used in direct contact between individuals. The exact communicative function of skin glands is not known (Scott and Fuller, 1965; Fox, 1971; Thorne, 1992).

Cats

Visual communication

Like the dog, the cat expresses its visual signals mainly by the positions of the ears and tail, and by the body posture (Fig. 12.1). The ears can be moved either backwards while still erect, downwards towards the chin, or positioned partly backwards and partly downwards. Wild cats like the tiger or the lynx have dark markings on the back of the ear. In the backward ear position, these markings are seen from in front of the cat and make the ears more conspicuous. This ear position is thought to signal offensive aggression, i.e. the cat trying to manipulate the opponent to withdraw. By flattening the ears towards the chin, the ears and their offensive signals are hidden. This is used by defensive cats signalling acceptance of a subordinate status. Such cats can, however, defend themselves if the dominant cat attacks. This is frequently signalled simultaneously by an open mouth, perhaps also accompanied by hissing. The arousal of this cat is also evidenced by large pupils. If the cat experiences a conflict between offensive and defensive motivations, the ear signal may be intermediate (Fig. 12.1), providing the cat with a fine-tuned mechanism for expressing various relative strengths of these motivations. As staring eyes may signal offensive threat, cats frequently avoid looking at each other in the early stages of agonistic encounters.

Fig. 12.1. This cat expresses defensive aggression by his whole body. The ears are partly backwards and partly downwards, the eyes are almost closed to avoid damage, the paw is partly lifted to exhibit a certain propensity for its use, and the back is lowered.

While the ear signals can express short-term fluctuations in motivations, the body posture provides a more general signal about the relative motivations. The offensive cat stands erect with the tail vertically downwards, now and then looking to the side and showing its body as large as possible. The defensive cat may instead crouch down, hiding its head and tail. All kinds of intermediate postures may be seen. A cat arching its back, staying on stiff legs, piloerecting and with the tail partly or completely raised, expresses a strong conflict between offensive and defensive tendencies. This may, for example, be seen in cats defending an important resource, such as its offspring, where flight is not an option, or a cat being threatened by a dog. In other contexts, a tail-up signal may show the cat's intention to interact amicably with another cat or a human.

Acoustic communication

The acoustic repertoire of the cat is impressive. Aggressive cats may growl, yowl, or snarl, by which the mouth is held open in one position, or howl by opening and gradually closing the mouth. Defensive cats may hiss or spit. Much more pleasant is the purr and the trill or chirrup (sometimes written as 'mhrn). A purring cat signals a high motivation for contact or solicitation of care. Therefore purring may also be heard in cats experiencing illness or pain. The chirrup is a greeting signal, or a call if it starts abruptly. This may sometimes serve as the start of a miaow, which may also be a greeting signal, perhaps at a slightly larger distance. Towards its mother or a human, the cat may vary the miaow enormously. Based on extensive studies by Mildred Moelk (1944), the cat's motivation or manipulative intent is interpreted in relation to which of the four main 'letters' of the miaow are most pronounced (Fig. 12.2). A marked '*m*', like the chirrup, signals contact motivation. A long '*i*' signals distress or pain, a long '*a*' signals a demand for a resource, while a long '*ow*' signals a frustrated demand. Because of the marked individual variation, any interpretation must be adjusted by knowledge of the cat's inclination to use these signals.

Olfactory communication

Precise knowledge about olfactory communication is difficult to obtain. Urination can be performed either by squatting (by bent hind legs) or by spraying (in upright position). Both seem to convey some individual information, although the latter is specialized for scent marking and includes secretions from the anal gland. Spraying is used by both males and females. At high population densities, mainly the dominant males spray. Cats do not scent mark the edges of their territory. Instead, urine marks may be found along tracks near their core area. Other cats do not seem to be deterred while sniffing the scent, often using the so-called flehmen behaviour, whereby the upper lip is curled and retracted. By sniffing, they may get information about the probability of encountering the scent-depositing individual. Females also scent mark to announce

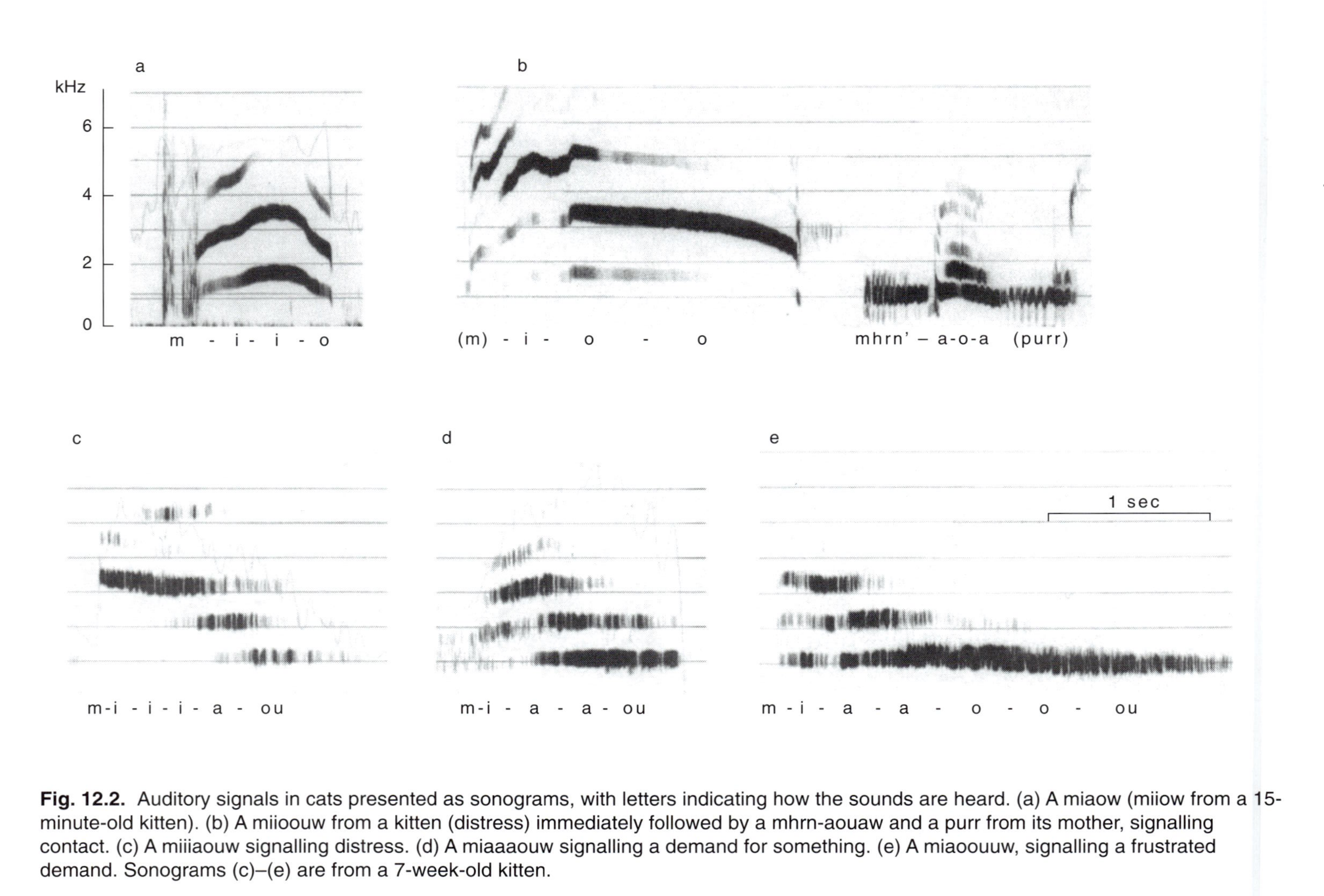

Fig. 12.2. Auditory signals in cats presented as sonograms, with letters indicating how the sounds are heard. (a) A miaow (miiow from a 15-minute-old kitten). (b) A miioouw from a kitten (distress) immediately followed by a mhrn-aouaw and a purr from its mother, signalling contact. (c) A miiiaouw signalling distress. (d) A miaaaouw signalling a demand for something. (e) A miaoouuw, signalling a frustrated demand. Sonograms (c)–(e) are from a 7-week-old kitten.

their oestrus. The cat rubbing its mouth, chin, flank or tail against an object, or scratching it with its claws, deposits saliva or secretions from skin glands. The functions may be similar to scent marking, as it is most common in intact males and oestrus females.

Tactile communication

Rubbing and licking may also be performed against other cats or humans. This may be a way of developing a common group odour, as well as a tactile signal expressing social bonds (Leyhausen, 1979; Turner and Bateson, 2000, Chapter 5).

Predatory Behaviour

Dogs

Wolves are primarily predators which hunt prey of all sizes – from the largest ungulate down to hares, rabbits and rodents. They also eat fruit and berries. When hunting the largest deer types, wolves hunt in packs and all attack the prey, weakening it by repeated bites in order to bring it down. Wolves can also muster a group of prey over extended periods of time.

For hundreds of years, if not thousands, humans have selected dogs for specific behaviours based on wolves' hunting behaviour and senses. Selection has favoured the initial components of hunting behaviour and selected against the final component. This form of selection has led, for example, to greyhounds which hunt primarily using vision, scent dogs specialized in tracking, sheep dogs and pointers, to name a few. Despite these significant differences between the various breeds, the main effect of selection has been a reduction or an increase of an already existing behavioural pattern, rather than the creation of new characteristics (Wills and Simpson, 1994).

Cats

Cats eagerly hunt small birds and small rodents like mice and voles, even if they are well fed by their owners. In Britain, it is estimated that feral cats may consume up to 24% of the annual production of field voles, although this figure is rather uncertain. As rats may defend themselves fiercely, cats usually restrict their interest in this species to young rats. If rat populations are reduced by other means, cats can suppress a subsequent rise in rat numbers. This may be particularly evident if cats are removed from the area, e.g. by prohibiting the keeping of pet cats. Also rabbits and young hares may be predated. On a few local sites, e.g. on islands, cats are reported to harm bird populations. Apart from this, cats have not been documented to be a major threat to bird populations.

Cats use their acute hearing and vision for detecting prey. Ultrasound communication signals (>20 kHz) in mice can attract the cat's interest. When the cat sees a small, moving object, its predatory behaviour is further stimulated. That is why the cat readily exhibits object play, an important aspect of the cat's behavioural development thought to provide the cat with hunting experience. When hunting mice the cat may wait in ambush, often extremely patiently. Bird hunting more often requires stalking, slow approach, and at last a sudden, fast and short gallop to catch the prey, sometimes including a jump into the air as the bird flies up. If the cat is not hungry, the killing bite may be inhibited and the prey may escape. The moving prey may then stimulate a new attack, resulting in the well-known 'playing with their prey' (Leyhausen, 1979; Dyczkowski and Yalden, 1998; Turner and Bateson, 2000, Chapter 8).

Mating Behaviour

Dogs

In dogs, oestrus generally occurs twice a year, after reaching sexual maturity at 6–14 months of age. Females in a group will generally synchronize their heat periods, which last about 21 days. The onset of heat is indicated by a swollen vulva, after which an opaque liquid is discharged, which gradually changes to blood-stained fluid. During the first 10 days females frequently urinate and deposit odours which can attract males from great distances, although the females at this stage show little interest in mating. Between days 10 and 15, females will allow courting and mounting by the male. However, some females will accept a male until day 21. Males are always ready to mate, but only display sexual interest in a female during her heat. Prior to copulation, dogs generally display courtship behaviour, such as play-fighting and mutual licking of genitals and ears. The female will repeatedly display a mating posture with her tail clearly to the side of her genitals. Despite being ready to mate, some females display aggression towards non-acceptable males.

During mating, the male mounts the female, holding her ribcage with his forepaws and bringing his penis close to her vulva to enable penetration. The genitals of the canine family are unique; they contain a structure known as the *bulbus glandis*. Following ejaculation, which occurs shortly after penetration, the *bulbus glandis* swells and locks the two individuals together. This allows the male to dismount the female by lifting his hind legs over the females back. The two will stay locked together for 10–30 minutes with their hindquarters facing each other. Following the reduction of the *bulbus glandis*, the individuals separate, and the male lies down and licks his penis until it retreats into its sheath.

Cats

The home ranges of male cats are particularly large during the mating season, in which it overlaps with home ranges of several females. Female cats are seasonally polyoestrous, coming into heat at intervals of a few weeks. Ovulation is induced after several copulations, which may total 15–20 per day over 4–5 days. These may be adaptations to maximize male competition for females. Increased scent marking by the female during oestrus, and sometimes interfering to break up a 'locked' dominance situation between males, may add to this. The courtship behaviour may be long lasting, but copulation takes only a few seconds. Female cats are often reported to prefer a familiar male to an unfamiliar one, even if the latter is the dominant one (Bradshaw, 1992; Turner and Bateson, 2000, Chapter 7).

Parturition and Parental Behaviour

Dogs

Whelping takes place approximately 63 days after mating. As term approaches, the female generally becomes restless and attempts to find a dark and sheltered place to prepare a nest. During the final day before parturition, many females refuse food and pant. This stage can last for more than 12 hours. The birth of the pup usually takes a few minutes, the interval between pups being 20 minutes to 2 hours. The moment a pup is born, the female will lick it, releasing it from the birth sac and enabling it to take its first breath. The female then severs the umbilical cord and swallows the placenta, licks the puppy dry, lies down and encourages it to nurse. The mother spends most of her time with her pups during the first days, while the pups may suckle for up to 30% of the time. While the puppies suckle, their mother cleans them, and stimulates them to urinate and defecate by licking their ano-genital regions. She swallows all their excretions, keeping the nest box clean and odour-free. As the pups grow and develop, feeding no longer dominates their life, and the female will leave them alone for increasingly longer intervals. At 3 weeks of age, some mothers returning to their pups will regurgitate food for them, as well as allowing them to nurse. The female will lay down to nurse less often; instead they suckle as she stands. Weaning is in progress. At 5 weeks of age, the mother may growl and show her teeth at her pups when they try to suckle. She may even snap at the pups. The pups often roll over on their backs, whining. These methods may seem harsh, but the pups are seldom or never injured, and after a short while they learn to stay away.

Cats

The African wild cat, *Felis silvestris libyca*, is monogamous, which may be necessary to gain enough resources and food for raising kittens in the semi-desert conditions in North Africa. Reminiscence of this monogamy may explain the several anecdotes of male domestic cats taking part in caring for kittens. More typically, the female cat does this alone. Parturition in domestic cats is usually extremely easy, without the birth problems often seen in other domesticated species. Some inexperienced mothers may hesitate to initiate normal maternal behaviour and may generally be less effective mothers. During the first days after parturition, the mother cat may nurse her kittens for as much as 6–8 hours per 24 hours. In group-living females communal rearing may occur, the females nursing any of the kittens in the group. After 3 weeks, suckling is gradually less frequently initiated by the mother and more and more by the kitten. From the fourth to fifth week after birth mothers bring home prey to their kittens, first dead prey and later live prey on which the kittens can practice prey catching.

Development of Behaviour

In both dogs and cats, sensory abilities slowly mature during the first 3–4 weeks of life. In kittens, the ear canal opens at 5 days and eyes on average at 9 days (range 2–16 days). From 3 weeks, skills in locomotor behaviour develop rapidly. Social relations develop during a sensitive period for socialization, which is roughly 3–10 weeks in both dogs and cats. In order to be completely tame, dogs and cats need positive contact with humans during this period, the most effective period for cats being 2–7 weeks. Social play and object play predominate from the second month of age. This may be important for refining the use and interpretation of communication signals, as well as hunting skills. The social bond between littermates seems to persist during the juvenile period and may endure throughout their lives. Although cats may be sexually mature at 6–9 months of age, males may not show a complete social behaviour before about 2 years of age. In the following, behavioural ontogeny of dogs will be treated in more detail. The progression in cats is not very different.

Dogs

New-born canid cubs are blind, deaf and completely dependent on their mother. As the physical development of the cubs progresses, they become more independent and more aware of their surroundings. Development is generally divided into four periods; the neonatal period, the transition period, the socialization period and the juvenile period (Fig. 12.3). In addition, the prenatal period may be included.

The prenatal period

The prenatal period has often been neglected in canids, with the exception of studies of farmed foxes which, in line with extensive studies on rodents, indicate long-term effects of the uterine environment on behavioural development. Females exposed to stress during pregnancy may give birth to young that are comparatively more emotional, reactive and fearful in later life. This is related to increased activity of stress mechanisms in the offspring.

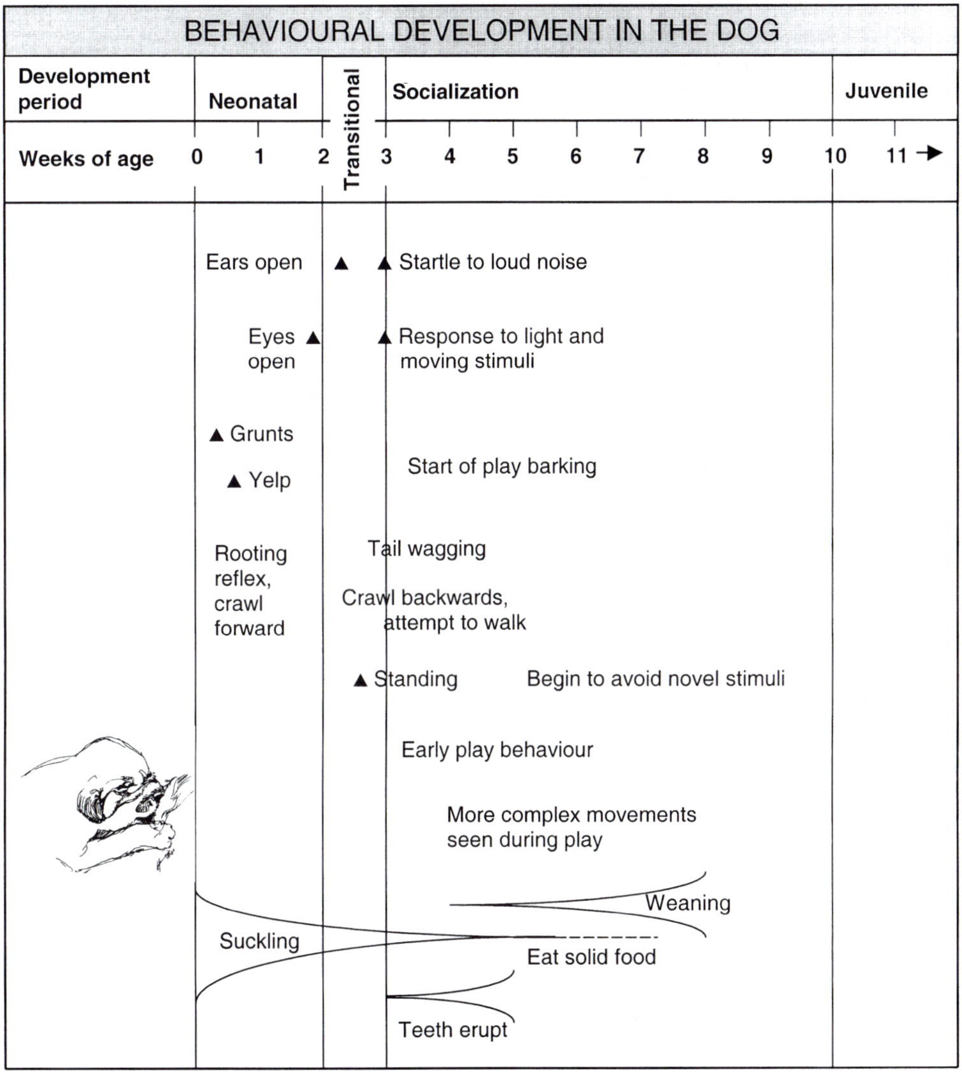

BEHAVIOURAL DEVELOPMENT IN THE DOG

Development period	Neonatal	Transitional	Socialization	Juvenile
Weeks of age	0 1 2		3 4 5 6 7 8 9	10 11 →

Ears open ▲ ▲ Startle to loud noise

Eyes open ▲ ▲ Response to light and moving stimuli

▲ Grunts

▲ Yelp

Start of play barking

Tail wagging

Rooting reflex, crawl forward

Crawl backwards, attempt to walk

▲ Standing

Begin to avoid novel stimuli

Early play behaviour

More complex movements seen during play

Weaning

Suckling

Eat solid food

Teeth erupt

Fig. 12.3. Major steps in sensory and behavioural development in the dog.

The neonatal period (0–14 days)

Pups are entirely dependent on their mother during this period. This stage is dominated by suckling behaviour, excretion and sleep. The sensory apparatus is poorly developed. The pup has no sight or hearing, although it reacts to tactile stimuli and possibly smell. The pup crawls slowly, throwing its head from side to side, often whining or yelping as it moves. However, if rolled on its back, it will struggle to turn over again. Pups will crawl, whine or squeal when exposed to hunger, cold or pain. Pups begin suckling when coming into contact with their mother's teat. In the most complete form of this pattern of behaviour, the puppy pushes on the teat with alternate forepaws, occasionally pulling back its head, bracing with its forepaws and pushing with its hind legs. This activity possibly stimulates lactation. After suckling, the pup falls asleep. Previous theories supposed that influences during this period would have no lasting effect, as the brain was thought to be poorly developed. However, pups subjected to early handling can have a greater ability to withstand stress in later life. In addition, wolf cubs handled by humans from birth exhibit friendlier behaviour towards humans later on than those handled after 15 days of age.

The transition period (14–21 days)

This is a period of change, where behaviour patterns adapted for neonatal life are decreased or abandoned, and the characteristic pattern of adult behaviour starts to appear. The onset of the period is marked by eye-opening at approximately 13 days of age, and ends when the ear channels open at 18–20 days. During the transition period, the puppy begins to orientate itself with its surroundings, it exhibits the ability to crawl backwards and forwards, and eventually rises to walk. The pup defecates and urinates without the tactile stimulation by its mother. The first attempts at play fighting and tail wagging are seen. The vocal repertoire increases and the pup will whine not only when cold or hungry, but also in an unfamiliar environment. By 3 weeks of age, teeth appear and puppies begin to bite and chew, clumsily, in preparation for the adult form of nutrition. Pups' learning abilities develop throughout this period, but do not reach a level comparable to older individuals until 4–5 weeks of age. By the end of the transition period, the pup is ready to progress to a stage in which it is extremely sensitive: the period of socialization.

The socialization period (3–10 weeks)

During the period of socialization, the puppy begins to show the majority of adult behaviour patterns, albeit in playful form (Fig. 12.4). It starts to respond to the sight and sound of other animals and humans at a distance away, inspecting humans and often wagging its tail. The first signs of fear appear. Aged 3–4 weeks, the pups start following each other and from

5 weeks may react to different stimuli as a group. They leave the nest for urination and defecation. The first signs of agonistic behaviour are revealed, and some breeds may even show the first signs of aggressive attack.

Social bonds with other pups, and with the mother and humans, are developed during the socialization period. Puppies that do not receive social contact with humans in this period will show fear towards humans at a later stage. The peak period of socialization is 4–8 weeks of age. Socialization is controlled by two central motivational systems: the pup's contact-seeking behaviour and fear. Contact seeking increases from 3 to 5 weeks of age and decreases thereafter. The puppies' level of fear, which rises continuously after 5 weeks, causes the decline in approach behaviour. A characteristic of

Fig. 12.4. Early play behaviour in pups with elements from an aggressive interaction. (Photographs by Jan Kolpus.)

silver foxes bred for tameness over 30 years in Russia was the delay in the closing stage of the socialization period from 6 to 9 weeks of age. This was associated with a postponed development of fear and aggressive reactions. The primary effect of this selection was not a genetically tame individual, rather an individual that could more easily form social bonds with people, through a significant increase of the duration of the socialization period. The dog has come even further in this process.

During the socialization period, pups are generally sensitive and susceptible towards different types of stimuli. Pups with broader experiences will be more capable of coping with various challenges later in life than pups brought up in a low-stimulus environment.

The juvenile period

Puppies grow rapidly during the first part of this period and are almost fully grown by 8 months of age. They lose their milk teeth at approximately 5 months, and these are quickly replaced by strong permanent teeth. During the juvenile phase, the basic behaviour patterns of pups do not change significantly, although there is a gradual improvement in motor skills. Puppies gradually learn the implications of their behaviour and are able to determine which behaviours are appropriate in specific situations. The most significant behavioural changes occur during sexual maturation between 6 and 14 months. Male dogs begin to lift one hind leg when urinating. Dogs of both sexes will often mark their social status towards other dogs, in some situations also towards their owners. Despite females being sexually mature and capable of reproduction, they are not yet sufficiently mentally mature to mate and raise pups (Scott and Fuller, 1965).

Dogs and Cats Interacting with Humans

Humans can form strong social bonds with dogs and cats. It is documented that this kind of frequent contact may lead to positive emotions and mental relaxation in many humans. This is associated with a number of beneficial effects on the mental and physical health of humans. Communicating with and caring for a pet may promote social behaviour and non-verbal communication in young children. In addition to providing companionship, pet keeping is associated with reductions in blood pressure and other risk factors for cardiac diseases, as well as reduced incidences of psychosomatic diseases and depressions. Animal-assisted therapy is a term used for more systematic utilization of dogs and cats, or other animals, in order to improve the mental and physical health, particularly of psychiatric or geriatric patients (Fig. 12.5). In other cases, those engaged in visitors' programmes in hospitals may bring a dog. In some places special animals are bred for this purpose. In order to ensure a harmonic relationship with the animal and positive effects on the human, the needs and welfare of the animals must be taken care of.

The sensory, physical and learning abilities of dogs are utilized for several purposes. Appropriately trained dogs may guide blind people, alert deaf persons about important sounds, perform particular tasks for physically disabled persons, and serve as police dogs, mine dogs, drug dogs and mountain rescue dogs (Fine, 2000).

Behavioural Problems

Behavioural problems in dogs and cats may be true behavioural disorders, abnormal behaviour patterns in a psychiatric sense that are not part of an ethogram of free-living populations, but more commonly they are normal behaviour patterns of the animals that are undesirable to humans. True disorders may develop if the owner provides the animal with inconsistent or unpredictable reinforcements, or if the welfare is severely challenged in other ways. A cat that is punished for defecating on the carpet long after the incidence has no idea why it is being punished. In the long run such cats may develop tenseness or fear when being close to human hands. Other problems such as dominance aggression, inappropriate urination or defecation, indoor scratching, barking, etc. may be a problem to humans, but not necessarily to the animals. It is usually much easier to prevent the development of such problems by careful consideration of the animal's needs and the way it is handled than to find efficient measures against them once established. However, the propensity for developing a certain behavioural problem may be affected by inheritance and may vary considerably between both dog and cat breeds.

Fig. 12.5. A pet cat may have positive effects on the mental health and well-being of humans. (Photograph by Alf Ramsfjell.)

Particular behavioural problems in dogs

While dogs were once kept primarily for specific uses related to farming, hunting and guarding in rural areas, they are now, to a great extent, kept as pets in noisy urban areas with high population densities of both people and dogs. These environmental changes place heavy demands on dogs' abilities to adapt. Although dog owners, to a certain degree, are aware of these problems, most will expect their pup to develop into a well-balanced and well-behaved individual. However, in some cases the trusting relationship is destroyed, as the dog develops a behavioural characteristic that the owner cannot control or accept; a behavioural problem. Occasionally euthanasia is the only option. Killing due to behavioural problems was the third most common cause of death for three dog breeds studied in Norway. The average age at euthanasia was 3.5–4 years of age. However, information from behavioural therapists indicates that owners lived with these behavioural problems for a long time – the dog is euthanized, on average, 2 years after the owner seeks help in dealing with behavioural problems.

Problems related to separation are among the most commonly reported behavioural problems, reported by over 20% of dog owners. These include vocalizations – barking, whining and howling; destruction – biting, chewing and scratching; and defecation and urination. The latter may indicate general anxiety, vocalizations may indicate attempts by the dog to restore contact with its owner, while destructive behaviour may, to some degree, be motivated by hunting behaviour.

Aggression-related problems often lead to euthanasia. Dog literature categorizes aggression in two ways: (i) according to the victim of the aggressive threats or attack – the owner or family members, strangers, or other dogs; or (ii) from a causal perspective, where the aggressive behaviour is correlated with the trigger-factor for the aggressive action – dominance, territorial, protection of property, aggression between males, aggression between females, fear-induced, pain-induced, punishment-induced, redirected aggression, maternal and instrumental aggression. These classification systems are often erroneously combined, so that, for example, aggression towards the owner or family members is classified as dominance aggression, although in many cases the aggression will be caused by fear, pain or punishment (defensive aggression). All types of defensive aggression, in which the dog uses aggression in an attempt to remove the source of the negative stimulus, will be enhanced through punishment. Repeated release of defensive aggression may lead to permanent aggression, despite the removal of the threatening or discomforting stimulus. Correct diagnosis of the cause of aggression is vital in reducing the frequency of the dog's aggressive behaviour.

Many dogs show, as do wolves, aggressive tendencies towards strangers or dogs within their territory. This is known as territorial aggression. Dogs can warn by barking or growling, and, sometimes, stronger forms of aggression. Territorial aggression usually appears between 1 and 3 years of age, and is a trait that has been enhanced through selection, especially in guard dogs.

Nervous and fearful responses to intense noise, strangers and unfamiliar situations are another source of behavioural problems in dogs. Fear or anxiety, as in humans and other animals, will strongly influence the dog's confidence in its everyday life. Generally, novelty, lack of predictability or controllability of a situation stimulate a fear reaction in the dog. Gradual introduction of novelty will reduce the possibility of a pup developing extreme anxiety reactions. However, the speed at which a pup adapts to new stimuli is highly dependent upon the individual.

As with all types of behaviour, behavioural problems will be the result of genetic predisposition combined with the dog's environment. Problems related to aggression, fear and separation all have high degrees of heritability. Therefore, it is possible to alter the response levels of these behaviours through selective breeding. In addition, prenatal experiences, influences during the socialization period and juvenile stage will all determine the dog's predisposition towards developing behavioural problems. However, the ultimate cause of behavioural problems still lies with the owners and their treatment of the dogs (Serpell, 1995).

Particular behavioural problems in cats

Scratching furniture or other inappropriate objects is the most common behavioural problem in domestic cats, according to a Norwegian survey of 831 cats. This occurred 'often' or 'very often' in 33% of non-pedigree cats, 27% of Siamese cats and 13% of Persians. Because the owner quickly approaches the scratching cat, this behaviour is unintentionally rewarded as a signal for calling the human. In general, such inappropriate contact signals must be ignored if they are to be counteracted. Urination outside the cat toilet and indoor urine marking were major problems, particularly in Siamese cats (18%), but only in 3–5% of Persian and non-pedigree cats. There are a number of reasons for these behaviours, e.g. related to isolation (contact marking) or social problems (territory marking; Heath, 1993). Various ways of stimulating the cat's self-confidence should be sought. Neutering the cat does not always solve the problem. Among non-pedigree cats, 17% show fear of unfamiliar humans, 16% exhibit aggression towards other cats, and 6% aggressively bite or scratch humans. The latter problems may be related either to (usually indoor) cats being understimulated, or to sudden fear during petting because the cat has experienced being handled unpleasantly. In pedigree cats, the highest heritabilities in the Norwegian study were found for playfulness, fear and behaviour towards unfamiliar persons. These traits could be considered when selecting breeding animals, in order to increase further the proportion of cat owners that are satisfied with their cat.

References

Bradshaw, J.W.S. (1992) *The Behaviour of the Domestic Cat.* CAB International, Wallingford, UK.

Clutton-Brock, J. (1999) *A Natural History of Domesticated Mammals*, 2nd edn. Cambridge University Press, Cambridge.

Dyczkowski, J. and Yalden, D.W. (1998) An estimate of the impact of predators on the British field vole *Microtus agrestis* population. *Mammal Review* 28, 165–184.

Fine, A. (2000) *Handbook on Animal-assisted Therapy: Theoretical Foundations and Guidelines for Practice.* Academic Press, San Diego.

Fox, M.W. (1971) *Behavior of Wolves, Dogs and Related Canids.* Harper & Row, New York.

Gittleman, J.L. (1989) *Carnivore Behaviour, Ecology, and Evolution.* Chapman & Hall, London.

Heath, S. (1993) *Why Does my Cat ...?* Souvenir Press, London.

Leyhausen, P. (1979) *Cat Behavior: the Predatory and Social Behavior of Domestic and Wild Cats.* Garland STPM Press, New York.

Moelk, M. (1944) Vocalizing in the house-cat; a phonetic and functional study. *American Journal of Psychology* 57, 184–205.

Morey, D.F. (1994) The early evolution of the domestic dog. *American Scientist* 82, 336–348.

Scott, J.P. and Fuller, J.L. (1965) *Dog Behaviour, the Genetic Basis.* University of Chicago Press, Chicago.

Serpell, J. (1995) *The Domestic Dog, its Evolution, Behaviour and Interaction with People.* Cambridge University Press, Cambridge.

Thorne, C. (1992) *The Waltham Book of Dog and Cat Behaviour.* Pergamon Press, Oxford.

Turner, D.C. and Bateson, P. (eds) (2000) *The Domestic Cat: the Biology of its Behaviour*, 2nd edn. Cambridge University Press, Cambridge.

Wills, J.M. and Simpson, K.W. (1994) *The Waltham Book of Clinical Nutrition of the Dog and Cat.* Elsevier Science, Oxford.

13 Behaviour of Rabbits and Rodents

David B. Morton

Origin and Domestication History

Rodents are the most common and diverse of mammals, comprising about 40% of all mammalian species (UFAW, 1999), their name being derived from the Latin *rodere*, to gnaw. Neither Bugs Bunny nor Mickey Mouse or Roland Rat reflect the fact that these animals are seen more as pests in most countries, in that they compete for the same food as humans (it has been calculated that they eat 42.5 million tonnes of food each year, i.e. some 5–10% of total production). These animals have been, and still are, important vectors of disease (they are involved in the transmission of more than 20 pathogens and are said to have killed more humans than all wars put together). In some cultures rodents are the objects of culinary delights (guinea-pigs and dormice), but they are also used as experimental animals, and as such have contributed enormously to our understanding, and treatment, of various diseases in humans. In some countries rodents and rabbits are increasingly being kept as pets and are cared for and protected by their owners in similar ways to the more common companion animals such as dogs and cats. Studying the behaviour of rodents and rabbits has become important from the point of view of controlling their population size and even exterminating them. But this knowledge has also been employed to keep them healthy, both physiologically and psychologically, when they are used in research laboratories and as pets. Their anatomy, physiology and genetics influence their behaviour, and indeed captivity may result in pathological conditions arising directly as a result of them not being able to carry out certain behaviours. This chapter will look at some of their behaviours as studied in research laboratories and, where appropriate, extrapolated to other situations.

Rodents

Rodents and rabbits are derived from the zoological orders of Rodentia and Lagomorpha and are found throughout the world (see Macdonald, 1995;

UFAW, 1999) and are distinct in several ways (see below). Rodents are defined by a continually growing (several millimetres/week) pair of incisors on upper and lower jaws, which differentially wear, ensuring a sharp edge to the teeth. Their well-developed head muscles enable them to crack open various seeds to obtain food, as well as to gnaw through relatively hard materials such as wood and even concrete. This means that the materials from which cages are made have to be chosen carefully and not be toxic in any way. Rodents tend to be omnivorous, but different species store food differently. Hamsters have cheek pouches, whereas others make stores or larders of food in their burrows. Their senses of smell, hearing, taste, touch and sight are well developed (see below, under signalling). They tend to be highly social animals, live in large groups with a hierarchy, and are able to reproduce rapidly and in large numbers (the young develop sexual maturity early and they have relatively short gestation periods). These strategies, when viewed from the high rates of predation that occur, would seem to help in promoting the fitness of the individuals.

The species commonly used within the order Rodentia, and which will be covered here, are the mouse (*Mus musculus*, of which there are 600 or more inbred and outbred strains), and the brown or Norwegian rat (*Rattus norvegicus*, again many inbred and outbred strains, but fewer than mice). Different strains of a species arose as a result of different breeding strategies (see UFAW, 1999), much in the same way as there are different 'breeds' of dogs. In order to reduce the effect of varying genetic make-up on research, scientists produced genetically 'identical' animals by brother–sister or parent–offspring matings for more than 20 generations, when it can be calculated that the genetic similarity would be more than 98.4%. In this way they also 'fix' certain desirable genetic characteristics as they became homozygous for that trait, e.g. to model a particular disease in humans. On the other hand, because of this homozygosity, undesirable traits also arose and so these animals often have smaller litters, and do not grow as large as they would do normally. Because genetic aspects were not always seen as important, many scientists and breeders deliberately outbreed their animals in order to ensure maximum genetic diversity as well as maximum productivity.

Other rodent species used in laboratories include the Mongolian gerbil (*Meriones unguiculatus*), the golden hamster (*Mesocricetus auratus*), the guinea-pig (*Cavia porcellus*), and the bank (*Clethrionomys glareolus*) and field voles (*Microtus agrestis*).

The importance of genetics should not be underestimated in determining behaviour, as it is well recognized that strain has an important influence on the differing responses of the same species of animal in similar or identical environments (van de Weerd *et al.*, 1994). Despite the fact that many species of rodents and rabbits have been kept in captivity for more than a century, they still retain many of their natural characteristics and instincts that evolved in their wild ancestors. A knowledge of their behaviours in the wild may therefore give us ideas about how to modify their captive environments to suit them better (Poole, 1997; Galef, 1999).

Rats

The laboratory rat (*Rattus norvegicus*) is a descendant of the wild rat that originated in Asia and reached Europe in the early 1700s, displacing the indigenous black rat (*Rattus rattus*). In the mid-1800s in the USA, these animals were caught on trading ships and whereas the brown rats went into the terrier pits (in this 'sport' they timed how long it took dogs to kill 100–200 rats), the white (albino) rats were used for showing or research (most notably at the Wistar Institute laboratories in Philadelphia, USA). Hooded rats are so called because of the dark fur over the back of the neck, and presumably were the early hybrids. The long, close association of rats with humans probably facilitated taming them.

Mice

The word mouse comes from the Sanskrit *mush*, meaning to steal, and both mice and rats have been well known for raiding grain stores for over 6000 years (Keeler, 1978). At certain times in some cultures mice have been revered and worshipped, and have even been used in medical concoctions (e.g. 10 new-born mice dissolved in olive oil and mixed with white and yellow *Artemisia* flowers was used for a variety of illnesses even until the 1930s). Like rats, mice were used in the early studies of anatomy and in research from the late 1700s onwards. However, during the 19th century the house mouse became part of the show trade, and it was cross-bred with mice brought in by British traders from Japan and China, and selected for coat and eye colour. The mouse was commonly used in mammalian genetic studies, inspired by Mendel's work, and their role in our early understanding of the influence of genetics in transplantation and other areas is now well recognized.

Other rodents

The hamster comprises several strains, the most common of which is the golden or Syrian hamster (*Mesocricetus auratus*), which has a grey belly and a reddish brown coat. Hamsters are mainly nocturnal and live naturally in burrows in dry, rocky or brushy steppe-type ground. They are thought to have originated in the Western world from a few pairs caught in Syria in 1930 (Murphy, 1985) and these are the ancestors of those used in the laboratory or as pets.

The guinea-pig comes from South America (Weir, 1974) where it is found in abundance in Argentina, Uruguay, Bolivia, Brazil and Peru, apparently first introduced by the Portuguese into Brazil. It neither comes from Guinea (west coast of Africa) or Guiana (north-east coast of South America) and nor is it a pig, so how did it get this name, a name that reminds us that it was used in research? The 'pig' part is thought to derive from its resemblance to a suckling pig, and 'guinea' as a result of the countries visited by the trading vessels in the South Americas and Africa. In the 16th century the Dutch introduced the guinea-pig into Europe, but it failed to gain acceptance as a source of food; however, it did spread as part of the pet and show animal trades. From there it was used in research in the 1930s, although Lavoisier had used these animals in 1780 to measure heat production.

Gerbils originated in Mongolia and China, are saltatorial (move in jumps or hops like kangaroos) and have a long balancing tail, long hind legs and short front legs. They have a ventral scent marking gland in both sexes, the size of which is sex-hormonally dependent. They are adapted to live in deserts and have the ability to concentrate urine, unlike other rodents, surviving for several weeks on the small amount of water found in relatively dry foods such as seeds.

Rabbits and hares

Lagomorphs (rabbits and hares) are characterized by fur that is usually long and soft and that covers their feet which, unlike those of rodents, provide a good grip with their strong claws. They have relatively large mobile ears (cooling and alerting to danger), high-set large eyes (alerting to danger at twilight, and all round-vision), and paired upper and lower incisors but, in addition, there is a second pair of peg-like incisors behind the first pair. Their long hind legs enable them to run away from danger.

Domestication of hares and rabbits probably started in Roman times, when they were used for food and reared in 'leporaria' or hare gardens (Thomson and Worden, 1956; McBride, 1988; Leach, 1989). In the 16th century, Queen Elizabeth I had a rabbit garden and Henry IV of France had a large enclosure that was used as a rabbit hunting area by women – it was an easy and safe hunting sport for them! The monasteries also maintained colonies for food and bred them for fur, and it is thought that the show trade based on fur colour started at about that time. Rabbits escaped from these areas into the wild, where the species has survived extremely effectively for centuries. For example, in 1859 a single pair of rabbits was released into Victoria in New South Wales, Australia and by 1890 the population was estimated at 20 million. There are over 50 strains today, but the emphasis has moved away from simply fur colour and quality, to selection for ear length (up to 25 cm) where the animals cannot support them – hence the name 'lop' meaning to hang loosely. They have also been selected for body weight, from less than 1 kg (Polish dwarf) to more than 8 kg (German giant), as well as growth rate and food conversion efficiency for meat production. Only the rabbit (*Oryctolagus cuniculus*) will be covered in this chapter.

Social Behaviour

Most rodent species live in social groups and may do so in captivity (Fig. 13.1); in fact keeping them singly results in detrimental outcomes, such as decreased immunocompetence and an increased tumour incidence.

While rodents, such as guinea-pigs, are fairly non-aggressive animals and live harmoniously together, Syrian and Chinese hamsters can be very aggressive and fighting can take place to the point where animals may be seriously injured and killed, even though they may mate first (note that in the wild the male would escape, but in captivity it is confined in a cage). However, it has been found that it is possible to keep littermates together provided they have not been separated from birth. In all species, the introduction of strange animals will prompt some form of aggressive behaviour, and so re-grouping or setting up new groups of animals has to be done carefully and, preferably, before they are sexually mature. They are also likely to want to choose their cage mates, as well as the type of space, but they can only do this if they are group housed.

A social group of rodents may be dominated by a single male, but in laboratories, where single-sex housing is practised and choices are limited for the animals, a different sort of hierarchy evolves. The dominant animal is often easy to recognize as, for example, in a cage of male mice, where it would be the one with little evidence of injury from fighting (injuries to the tail, rump and ear) or barbering (Figs 13.2 and 13.3). Barbering is when the dominant animal chews an area of coat, and often all the whiskers (vibrissae), of the sub-dominant animals, and it occurs in females as well as males.

Fig. 13.1. Rabbits and a guinea-pig, group housed with places to hide and a varied diet.

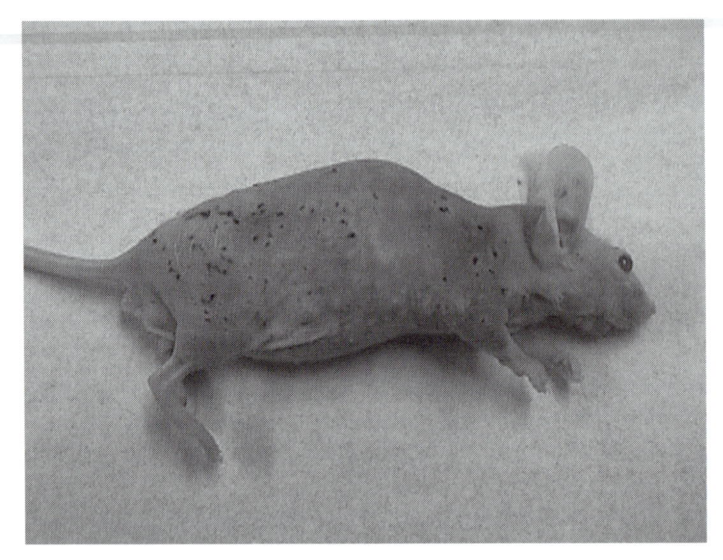

Fig. 13.2. Nude mouse found severely bitten overnight.

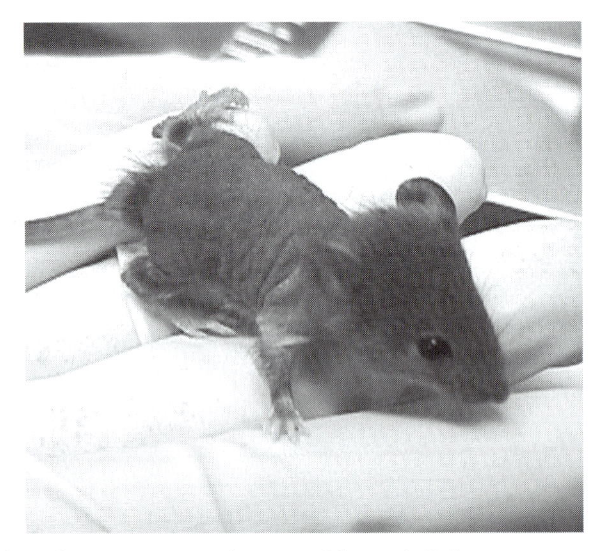

Fig. 13.3. Barbered mouse – note absence of fur and vibrissae.

Mice are notoriously difficult, unlike male or female rats, to keep in stable groups, i.e. with little or no aggression. They have a very well-developed sense of smell and so odour cues are very important; strains vary in their level of aggression. These odour cues stabilize the group, and when their cages are cleaned out, they immediately become aggressive until the cues are re-established (van Loo *et al.*, 2000). The provision of objects in which animals can hide is important to help them protect themselves.

Rats are fairly easy to house together in captivity, and even the introduction of strange animals is reasonably well tolerated. As with all species, the provision of shelters or retreats is an important escape provision at times. In the wild, rats form social groups around a dominant male with his harem, with subordinate males. They are sensitive to social isolation before weaning, and after they have been caged with others. Long-term isolation has been shown to affect their physiology (there is an increase in the weights of the adrenal and thyroid glands, but a decrease in spleen and thymus weights), behaviour and even their ability to problem solve. Single, caged animals are more aggressive and may be more active due to them carrying out stereotypic behaviours. Wistar rats kept in twos or threes show less stereotypic behaviour than single, caged animals. Animals seem also to benefit from being able to see, hear and smell adjacent animals, even if kept singly.

Gerbils live in very large groups in the wild, extending over several hundred square metres, and densities of more than 2500 gerbils per hectare have been recorded. They appear to have a strong need to burrow and their burrows may be found 1–2 m underground and extend for several tens of metres, with many openings to the surface and interconnections. Gerbils store their food and nurture their young in these burrows, and, as a consequence of the burrow depth providing insulation from heat and cold, they are able to survive extremes of temperature.

When rabbits are kept in groups, a dominance hierarchy is again set up and they may fight and seriously injure themselves, particularly males (often made worse if nearby females are in oestrus). If rabbits are neutered, then this abolishes aggressive behaviour in both males and females; in one such study, post-pubertal castration reduced aggression from 21% down to 0.7% (time budget analysis) almost immediately. When females are kept in groups they spend more that 75% of their time lying against another rabbit, and grooming each other (allogrooming), which would suggest that it is important for them to be with other rabbits and not kept singly.

Communication

All animals will squeal and vocalize in some way if they are hurt or frightened, but both rodents and lagomorphs may also freeze as an escape mechanism. In rabbits, vocalization can be a very penetrating sound and may put off a predator. Rabbits are also well known for thumping the ground as a danger warning to others and 'lookout' rabbits seem to be assigned for this task.

Chemical signals

Rodents and lagomorphs communicate by scent (scent glands and urine marking) and smell (used for identification of others, social status and state of oestrus). These scent glands may be in the head region (e.g. cheek glands of rabbits) or over the ventral abdominal area as in gerbils, and are frequently used to delineate territory, but this, of course, is very much reduced in captivity. Mice particularly use odour cues to identify each other and their status. Pheromones are important in mice from a reproductive view and can affect the rate of reproduction under certain circumstances, e.g. to delay maturation and implantation in adverse conditions such as overcrowding. Thus, mated female mice, when placed with a second male within 24 hours of the first mating, will abort implantation and be ready to mate with the second mate some 4 days later (Bruce effect). Pheromones have also been shown to synchronize the oestrous cycles of mice when a male is placed in a cage of females (Whitten effect). But note these may be artefacts due to abnormal groupings in captivity.

Interestingly, rats avoid foods that make them ill, and even pass this information on to others and so taste (or smell?) may also be well developed. This can make a population of rodents difficult to control by poisons.

Acoustic signals

Rodents are able to hear at levels outside the human range, notably into the ultrasound frequencies (they emit and detect ultrasound between 22 and 80 kHz). It is thought that this ultrasound can be stressful and induces 'abnormal behaviours' such as cannibalism (mothers eating their young) and stereotypies, as well as sudden noises (turning on a tap) evoking a startle response. Adult rats emit ultrasound during mating and during aggressive behaviour, and very young rat pups emit ultrasound when communicating with their mother or when left alone.

Tactile signals

Rodents' sense of touch is well developed through their vibrissae, which have a sensitivity of less than 90 μm – equivalent to the fingers of primates.

Vision

Albino animals, of which there are many strains in mice and rats, may find bright lights painful. They seek dim areas and try to hide from lights that humans would not find uncomfortable, such as 200–400 lux. Such levels will induce blindness in rats and should be decreased to 50 lux or so.

Foraging and Feeding

Both Rodentia and Lagomorphs show coprophagy (eat their own selected faecal pellets) that enables them to maximize the nutrient value of what they eat. This happens because on first pass through the gut, the large intestine breaks down the cellulose that is then further broken down in the small intestine to its constituent carbohydrates on the second pass.

Rodents are normally fed a pelleted chow in laboratories but, if permitted, like pet rodents, they will eat a broad variety of food, such as grain, dairy products, vegetables, fruit and various meat products. Water consumption will depend on the type of food being eaten; it will be less the higher the water content of the intake. Mice need to eat to maintain body temperature and they are able to survive in cold stores by building nests and eating sufficient quantities of food that the heat released in metabolism maintains their body temperature.

Rodents and rabbits forage naturally, but in the laboratory this behaviour is not necessary as they are fed *ad libitum*, which no doubt may be a cause of boredom. In the wild, food may be in short supply or be plentiful, but it will still take the animals time to ingest adequate amounts as it is far less energy dense than pelleted diets. Interestingly, when rats are given food in their hoppers and on the floor, they prefer to forage rather than take the same food from a full hopper. They also appear to prefer to work for their food, thus they will press a lever rather than eat food in front of them, and remove the husk from a sunflower seed rather than eat one already prepared. This behaviour is known as contra-freeloading (Inglis *et al.*, 1997) and has been observed in a variety of different species. It may be possible to enrich the diet of animals by providing them with a variety of more natural foods as well as the standard chow (Fig. 13.4).

Fig. 13.4. Flaked maize, wheat kernels, soaked and dried pellets and jelly (in the dish), add variety to a standard diet.

Rabbits eat at dawn and dusk, whereas rats eat 80–90% of their daily intake during the night, and mice eat during both the day and night. Many rodents store food in larders underground in the wild, but this is not possible in cages and they do not seem to create even primitive food stores (except perhaps hamsters).

Biological Rhythms

Clear diurnal rhythms, as well as some breeding seasonality, are apparent in rodents and rabbits. By and large, rodents are nocturnal and they have heightened periods of activity during the night with higher activity, heart rates and body temperature (no doubt all linked). Rats seem to have three activity periods at the beginning, middle and end of the night and take in 3–5 separate meals. Rabbits and guinea-pigs are crepuscular, i.e. are more active at dawn and dusk. Rabbits show most of their stereotypic behaviours during the night and not during the day.

Mating Behaviour

Sexual maturity usually occurs very early in life for rats, mice and other small rodents (some 6–8 weeks), but later for rabbits and guinea-pigs (3–6 months), and is strain dependent. Female rodents are spontaneous ovulators and are generally polyoestrus all year round, cycling every 4–5 days for the small rodents, but guinea-pigs have a 15–19-day oestrous cycle. Rabbits, on the other hand, are induced ovulators, i.e. mating induces ovulation, and will breed all year round, although they tend to have a lower fertility in the autumn. Female rodents will only accept a male when they are in oestrus, but may mate with several males, depending on the opportunities. However, as the period of oestrus lasts for less than 1 day, and mating often occurs at night, the chances of multiple matings with several males are much reduced, although multiple intromissions, but not ejaculations, by the same male are observed in some rodent species (e.g. hamster). Many rodents and rabbits show a post-partum oestrus (are in oestrus shortly after birth) and will mate successfully at that time. In laboratories a 12:12 hour day: night cycle will maintain normal reproduction in rodents. Mating usually occurs in the early hours of the morning, but rabbits will mate at any time. Mating can be confirmed through the presence of a vaginal plug (the product of accessory gland secretions in the ejaculate) that holds the semen in the vagina and uterus for 6–24 hours, after which it drops out. In gerbils, the presence of mature males and females appears to inhibit the sexual development of young gerbils, which may help prolong the existence of a family group.

Birth and Parental Behaviour

Gestation lasts around 21 days in mice and rats, 58–75 days in guinea-pigs, and 28–32 days in rabbits. Interestingly, the litter size and sex ratio of some rodents, e.g. mice, can be influenced by season, population density and food availability in the wild. Female rodents and rabbits build nests when they are about to give birth, and it is likely that this is a strong behavioural need for these animals. The nests of mice have been shown to have specific shapes (as in birds), and so the provision of suitable nesting materials may be important for pregnant females. The nest protects the young in certain environments (it provides cover and warmth) as the pups are unable to control their temperature until they are 2–3 weeks old. Birth normally takes place during the dark period, with rats, mice and rabbits tending to have large litters (7–16) of immature young. Guinea-pigs have only 2–4 mature pups which are well developed (precocial), with fur, and that are able to move shortly after birth. The female rat or mouse often walks round her cage to deliver the young and when born they are cleaned and then taken to the nest. Several minutes normally elapse between the birth of each pup, but if the dam is disturbed she may eat her young; this happens more commonly with primiparous females. It has been suggested that dominant male gerbils use cannibalism as a food source, but they usually do not eat their own offspring.

Rodents and rabbits have two lines of mammary glands on their ventral abdomen (10–14), whereas guinea-pigs have only two glands in the inguinal region. Rats and mice feed their young several times a day, roughly every 4–6 hours, whereas rabbits feed them only once a day. It is easy to see a stomach full of milk through the furless skin of neonates. In laboratories, the young are weaned at 3 weeks old, and this may happen naturally if a post-partum mating has taken place, as the subsequent litter would be due at that time. It is possible to cross-foster mice and rat pups, and it is more successful if done when the pups and the dam are at equivalent stages, but it is not unknown for lactating mothers just weaned to foster pups only a few days old.

Ontogeny and Development of Young

The young of rats, mice and rabbits are born with no fur, closed ear-flaps and eyes, and are unable to walk, whereas guinea-pig pups are precocial. Gerbils kept and reared under standard laboratory conditions with no access to cover develop quicker than those provided with shelter or tunnels, as evidenced by the age at which their eyes open (about a day earlier), their rate of growth, and time to reach sexual maturity; they are also less reactive to stimulation. Moreover, the adrenal glands of gerbils raised in open cages are lighter, but their reproductive organs and their pituitary glands are heavier than those raised with access to shelter.

Recent experiments have indicated that the nervous system of the young is not mature at birth because of the gradual myelination of nerves that takes place after birth and, more importantly perhaps, the growth of the descending inhibitory neurons. These descending axons synapse with other neurons in the spinal cord and modulate the passage of nerve impulses from noxious stimuli in the peripheral nociceptors (pain sensors). This may mean that very young animals are more sensitive to pain for the first 2–3 weeks of their lives than adult animals with a fully developed nervous system. Furthermore, the seemingly uncoordinated limb movements in the young are gradually refined and more precise as their proprioceptors develop and mature, as well as the myelination of the motor nerves.

Applied Problems

In laboratories, the design of animal cages has been centred mainly on considerations of human welfare, and animal health in so far as the cages are hygienic. Cages are made from materials that are durable, easy to clean, can be autoclaved (metal or plastic) and have wire mesh grid floors. They are shaped so that large numbers of cages can be placed on racks, or stacked for cleaning and storage, and can be easily handled (small, oblong boxes). They have a grid top and are sometimes transparent so that animals can be more easily observed. This system enables animals to be kept at artificially high stocking densities, often in single-sex groups. None of these characteristics provides for the psychological well-being of the animals, and contrasts with what animals choose in preference tests. One only has to consider the rich diversity of the wild to start to understand the different and restricted choices animals have when kept as pets or in the laboratory. In nature, rodents such as rats prefer to live near water in burrows some 50–70 mm in diameter and running for some 2–4 m, which end in nesting and food-storage areas. In laboratories they cannot burrow, cannot live near water (except near the water bottle), cannot run more than 50 cm, have restricted height for rearing and nowhere to store food. At worst, where there is no enrichment, they may live on simple gridded floors with the same diet and warm clean water from a bottle every day of their lives. Now consider for a moment the rich textures and sensory stimuli in nature – odours, vegetation, sounds of other animals, wind and water, and so on. Captivity is certainly much impoverished in comparison, even though it may be safer in some respects.

When kept in cages, all animals have to be protected from lying in their own excrement, and gridded floors were thought to be the answer, but in preference tests rats and mice choose to spend more time in cages with solid bottoms (Blom *et al.*, 1995; van de Weerd *et al.*, 1996). A variety of bedding substrates are used to absorb urine, such as corn cobs, sawdust, wood chips, wood shavings, paper, peat and so on, and the faecal pellets, although often passed in a defined area of the cage, are soon spread around the cage floor as animals move about. Preference tests have shown that rats prefer large particles as bedding, as they can manipulate it, and small particulate dusty bedding is avoided (it can also cause disease) (Blom *et al.*, 1996; van de Weerd *et al.*, 1997). Adding substrates such as sawdust, hay or straw, and paper into which they can burrow, hide, play and manipulate (make nests), makes for a more stimulating environment and animals choose to be in cages with these enrichments (Jennings *et al.*, 1998; Patterson-Kane, 2001; Fig. 13.5).

Stereotypic behaviour is extremely common in laboratory animals and pet animals if they are kept in a similar state of confinement. Mice show circling particularly, on the grid lids, but they also dig in corners – a behaviour commonly observed in gerbils and rabbits. Gerbils may even wear away the sides of their cage as they dig so hard and for so long. Voles show weaving, jumping and somersaulting, as well as sustained gnawing of the wire mesh of the cage. Such behaviours may damage the animals concerned and can interfere with care of the offspring. These stereotypic behaviours seem to be determined, and are certainly influenced by, centrally acting neuropeptides such as serotonin and dopamine, and can be modulated by drugs that compete with or bind to these neuropeptide receptors (see, for example, Schoenecker and Heller, 2001). Whether these types of behaviour are attempts to escape, or are induced by boredom or frustration from not being able to burrow and create tunnels, is currently being investigated. The time spent carrying out these behaviours has been reduced by increasing the cage size,

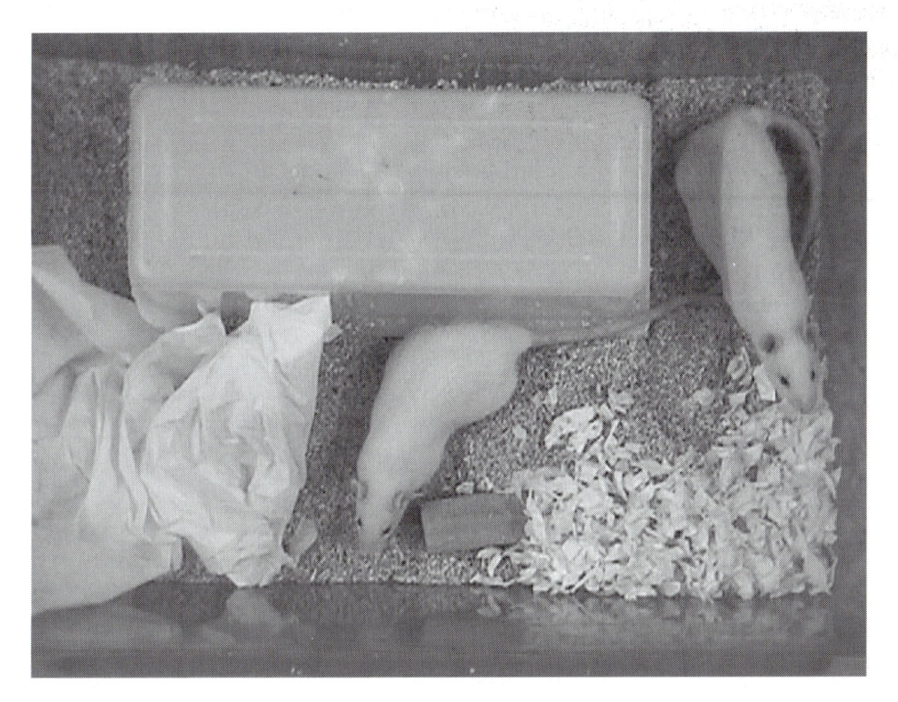

Fig. 13.5. Enrichment for laboratory rats – wood block, paper and shelter.

providing more hay or straw, by adding furniture such as plastic pipes and guttering, cardboard tubes and boxes (Scharmenn, 1991), and by providing preformed tunnels for gerbils.

The type of nest boxes chosen by mice and rats also varies, some strains favouring one with more than one exit hole, or different constructions or opacity. Rats spent 80% of their daytime in a tube as a shelter or resting place, and it may also protect them from the light. Animals kept in 'enriched' as opposed to 'barren' environments showed less anxiety in open field tests and explored more. Rats and mice appear to like to have things to chew, such as wooden blocks, and they play with other items, such as glass marbles and table tennis balls, as well, of course, with each other.

It has been shown that enriched and impoverished environments have substantial effects on brain development in rats and mice (see, for example, Kempermann *et al.*, 1997). When raised under non-barren environments their brains are heavier and they have increased neuronal branching, synapse densities, and astrocytes in various regions. This is seen functionally as these animals having better memory, learning ability, problem-solving ability, brain responses to undernutrition and old age, recovery from brain trauma and increased benzodiazepine receptor binding than those kept under standard conditions. This raises key questions about whether studies involving memory, or even perhaps 'normal' physiological research, can really give sensible and reliable results when conducted on animals that have been raised in such impoverished environments (Dean, 1999).

In addition, poor, unstimulating environments predispose to animals damaging each other more than if they are kept in so-called enriched conditions. Examples of enrichments include placing food, such as grain, in the sawdust on the floor so they are able to forage, providing them with objects in which animals can take cover and hide (pipes or boxes), as well as paper to shred and build nests. All these can help prevent serious injury from dominance behaviour (Fig. 13.6).

Rabbits show high levels of stereotypic behaviour when kept in cages on their own, which unfortunately is still common practice. Single housing results in behaviours such as weaving, chewing and excessive grooming, the latter predisposing them to denuding areas of their bodies and developing hair balls (trichobezoars). It seems that they become bored and will also chew excessively at their cage to the point where they may gnaw at the bars until they are bitten through and fractured, and in some cases damaging their teeth in the process. Rabbits are notorious for playing with their water bottles and emptying them overnight, and for trying to dig at the back of the cage (scratches can often be seen on the floor and sides of their cages). These stereotypic behaviours (which can occupy up to 25% of their time) are either simply not seen when rabbits are kept in groups (e.g. weaving), or they form part of their normal behavioural repertoire, but are only carried out to a limited 'normal' degree (e.g. licking as part of grooming).

There are many suggestions for cage size allowances for rodents and rabbits in various countries, and they are based on body weight. Surely, this concept is basically flawed and the housing of rabbits exemplifies the point. Young growing animals are more active than older and often fat, animals. However, because young rabbits weigh little, their space allowance is small, so small in fact that they are unable to carry out many natural behaviours (not seen in older animals) such as jumping, gambolling, and frisky hopping (jumping and twisting in the air). Moreover, as young animals are still growing, it may well interfere with normal tissue structure and development. This small space must frustrate their natural motivation more than older animals, who naturally exhibit less active behaviours. Should not more space (and height) be given to younger than older rabbits?

There is also a health cost to the animals on being reared in barren environments, especially if they are also markedly confined, like the rabbit. In addition to rabbits being more prone to developing hair balls, their caging predisposes them to being caged above their excrement for long periods and the ammonia fumes from their urine lead to irritation of their nasal mucosa and conjunctiva. The inability to move freely leads to thinning of the long bones and vertebrae, which makes these bones more likely to fracture when the animals are handled or when they are able to exercise freely.

Environmental enrichment techniques that appear to be the most successful for rabbits are those which provide the animals with the opportunity to express more of the range of species-typical behaviours. Perhaps the best form of manipulanda is hay, which has been shown to reduce the time they spend carrying out stereotypic behaviour. Rabbits will interact with other objects, which they chew and gnaw, such as plastic toys, empty cardboard boxes, food items to gnaw on (e.g. wooden or hay blocks), soda cans, wiffle balls, sections of PVC pipe, balls, and suspended metallic items. Objects suspended in a cage induce postures akin to the natural rearing position of rabbits, as well

Fig. 13.6. An example of enrichment in a mouse cage – wooden block, paper and shelter.

as play and investigational behaviours. It is interesting to note that some of the newer rabbit cages are larger than before, and are made from plastic, sometimes with a nest-box that can also be used as a 'lookout' shelf. Plastic, as well as being lighter than metal, also has a significant effect on their behaviour, as they spend more time lying stretched out (the plastic feels warmer?), carry out a broader repertoire of behaviours (more space), and social contact is permitted as more than one rabbit can now be kept in a cage.

When previously grouped rabbits are placed in cages for scientific studies, or occasionally to break up fighting, or when they are sick, it has been observed that they immediately start to show stereotypic behaviours that increase with time (from 0% in pens to 12.7% in cages), as well as an increase in behaviours indicative of boredom (lying motionless facing the back of the cage). Furthermore, subdominant animals showed more stereotypic behaviours than dominant ones.

Concluding Remarks

It is obvious that, in keeping with many other animal species, rodents and lagomorphs are no different in their requirement for an environment that is stimulating in various ways, and that depriving them of such conditions results not only in them becoming bored and exhibiting stereotypies, but also affects their natural development. The behaviour of these sorts of species may be more difficult to study than in other larger species, as their speed of movement, our lack of familiarity with them as individuals, and our attitude towards them as pests, will all influence how we care for these animals. This chapter has highlighted the choices that rodents and rabbits make for various environments, the impact of not meeting their needs, and questions as to whether keeping these animals for research in such barren environments may even be counter-productive and lead to obtaining misleading data.

References and Further Reading

Bayne, K.A.L., Mench, J.A., Beaver, B.V. and Morton, D.B. (2001) Laboratory animal behavior. In: *Laboratory Animal Science ACLAM series*, 2nd edn. Academic Press, London.

Blom, H.J.M., van Tintelen, G., Baumans V., van den Broek, J. and Beynen, A.C. (1995) Development and application of a preference test system to evaluate housing conditions for laboratory rats. *Applied Animal Behaviour Science* 43, 279–290.

Blom, H.J.M., van Tintelen, G., van Vorstenbosch, C.J.A.H., Baumans, V. and Beynen, A.C. (1996) Preferences of mice and rats for types of bedding material. *Laboratory Animals* 30, 234–244.

Dean, S. (1999) Environmental enrichment of laboratory animals used on regulatory toxicology studies. *Laboratory Animals* 33, 309–327.

Galef, B.G. Jr (1999) Environmental enrichment for laboratory rodents: animal welfare and the methods of science. *Journal of Applied Animal Welfare Science* 2, 267–280.

Inglis, I.R., Forkman, B. and Lazarus, J. (1997) Free food or earned food? A review and a fuzzy model of contrafreeloading. *Animal Behaviour* 53, 1171–1191.

Jennings, M., Batchelor, G., Brain, P.F., Dick, A., Elliot, H., Francis, R.J., Hubrecht, R.C., Hurst, J.L., Morton, D.B., Peters, A.G., Raymond, R., Sales, G.D., Sherwin, C. and West, C. (1998) Refining rodent husbandry: the mouse. *Laboratory Animals* 32, 233–259.

Keeler, C.E. (1978) In: Morse, H.C. III (ed.) *Origins of Inbred Mice*. Academic Press, New York, p. 181.

Kempermann, G., Kuhn, H.G. and Gage, F.H. (1997) More hippocampal neurones in adult mice living in an enriched environment. *Nature* 386, 493–495.

Leach, M. (1989) *The Rabbit*. Shire Natural History.

Macdonald, D. (ed.) (1995) *The Encyclopedia of Mammals*. Andromeda, Oxford.

McBride, A. (1988) *Rabbits and Hares*. Whittet Books.

Murphy, M. (1985) History of the capture and domestication of the Syrian golden hamster. In: Siegal, H.J. (ed.) *The Hamster*. Plenum, New York, pp. 3–20.

Patterson-Kane, E.G. (2001) Environmental enrichment for laboratory rats: a review. *Animal Technology* 52, 77–84.

Poole, T. (1997) Identifying the behavioural needs of zoo mammals and providing appropriate captive environments. *RATEL* 24(6), 200–211.

Richter, C.P. (1968) Experiences of a reluctant rat-catcher. The common Norway rat – friend or foe? *Proceedings of the American Philosophical Society* 112, 403–415.

Scharmann, W. (1991) Improved housing of mice, rats and guinea-pigs: a contribution to the refinement of animal experiments. *ATLA* 19, 108–114.

Schoenecker, B. and Heller, K.E. (2001) The involvement of dopamine (DA) and serotonin (5-HT) in stress-induced stereotypies in bank voles (*Clethrionomys glareolus*). *Applied Animal Behaviour Science* 73, 311–319.

Thomson, H.V. and Worden, A.N. (1956) *The Rabbit*. Collins, London.

Toates, F. (1995) *Stress. Conceptual and Biological Aspects*. John Wiley & Sons, Chichester.

UFAW (1999) *Universities Federation for Animal Welfare Handbook on the Care and Management of Laboratory Animals*, 7th edn, Vol. I. Blackwell Science, Oxford.

Van de Weerd, H., Baumans, V., Koolhaus, J. and van Zutphen, L. (1994) Strain specific behavioural response to environmental enrichment in the mouse. *Journal of Experimental Animal Science* 36, 117–127.

Van de Weerd, H.A., van den Broek, F.A.R. and Baumans, V. (1996) Preference for different types of flooring in two rat strains. *Applied Animal Behaviour Science* 46, 251–261.

Van de Weerd, H.A., van Loo, P.L.P., van Zutphen, L.F.M., Koolhaas, J.M. and Baumans, V. (1997) Preferences for nesting material as environmental enrichment for laboratory mice. *Laboratory Animals* 31, 133–143.

Van Loo, P.L.P., Kruitwagen, C.L.J.J., van Zutphen, L.F.M., Koolhaas, J.M. and Baumans, V. (2000) Modulation of aggression in male mice: influence of cage cleaning regime and scent marks. *Animal Welfare* 9, 281–295.

Weir, B.J. (1974) Notes on the origin of the domestic guinea-pig. In: Rowlands, I.W. and Weir, B.J. (eds) *The Biology of Hystrichomorph Rodents*, Symposia of the Zoological Society of London, No. 34. Academic Press, London, pp. 437–446.

Index

Browse
Read
and Buy
www.cabi.org/bookshop
ANIMAL & VETERINARY SCIENCES
BIODIVERSITY CROP PROTECTION
HUMAN HEALTH NATURAL RESOURCES
ENVIRONMENT PLANT SCIENCES
SOCIAL SCIENCES
CABI Publishing
A division of CAB International
Online BOOK SHOP
Subjects
Search
Reading Room
Bargains
New Titles
Forthcoming
Crop Pollination by Bees
A DICTIONARY OF Entomology
G Gordh and D H Headrick
Principles of CATTLE PRODUCTION
C.J.C. Phillips
Seeds
THE ECOLOGY OF REGENERATION IN PLANT COMMUNITIES
Order & Pay Online!
MasterCard
VISA
AMERICAN EXPRESS
FULL DESCRIPTION
BUY THIS BOOK
BOOK OF THE MONTH
Tel: +44 (0)1491 832111 Fax: +44 (0)1491 829292